RUSSIAN TANKS OF WORLD WAR II
1939–1945

STEPHEN HART

amber
BOOKS

Reprinted in 2020, 2022

Published by Amber Books Ltd
United House
London N7 9DP
United Kingdom
www.amberbooks.co.uk
Instagram: amberbooksltd
Facebook: amberbooks
Twitter: @amberbooks
Pinterest: amberbooksltd

ISBN: 978-1-78274-475-7

Editor: Michael Spilling
Designer: Andrew Easton
Picture research: Terry Forshaw

Printed in Malaysia

Picture Credits:
Nik Cornish/Stavka: 8/9, 22/23, 32/33, 48/49, 64/65, 80/81
Ukrainian State Archive: 6, 7, 18, 21, 29, 35, 36, 40, 43, 57, 71, 77, 79, 108, 114, 120, 121, 125

All artworks © Oliver Missing (www.o5m6.de) except 44/45, 62/63 © Art-Tech/Aerospace and 104–107 © Amber Books

Contents

Introduction

During the course of the desperate 1941–1945 struggle on the Eastern Front, the Red Army utilized over 85 major tank and self-propelled gun (SPG) designs to defeat the Axis invader; all of these armoured vehicles are described in these pages.

The mighty Soviet tank force that by May 1945 had defeated the German-led Axis powers traced its origins back to the 1919–21 Russian Civil War, during which the fledgling Red Army first captured 24 Ex-'White' tanks. Subsequently, by 1929 the Red Army could field 340 T-18 light tanks, a domestically-produced modification of the French-designed Renault FT.

After creating a potent tank industrial base, during 1929–40 the Soviets produced thousands of T-26, T-37A, T-38 and T-40 light tanks, 8000 highly-mobile, lightly-gunned and -armoured BT-2, BT-5 and BT-7 'fast tanks', 500 T-24 and T-28 medium tanks and 80 lumbering T-35 heavy tanks. The Red Army's impressive paper strength, however, concealed many flaws, such as obsolescent, under-gunned and inadequately protected light and fast tanks, poor build quality, bad crew ergonomics, mediocre optics and communication deficiencies. However, between 1939 and early 1941 the Soviets introduced promising new designs, including the T-60 Light, T-34 medium and KV-1 heavy tanks, as well as the KV-2 self-propelled gun (SPG).

German onslaught

The apocalyptic onslaught of Operation 'Barbarossa', the June 1941 German invasion of the Soviet Union, decimated both the Red Army tank force and its manufacturing base. Amid heroic

A whitewashed T-34 Model 1941 advances slowly through the snow near Kharkov in January 1942.

eastward factory relocation efforts, the Soviets initially mass-produced existing tank models (however obsolescent) plus expedient stop-gaps (such as the T-18M light tank and the ZiS-30 SPG), alongside the potent T-34 and KV-1 tanks, simply to compensate for the catastrophic losses incurred. These desperate measures helped the Red Army halt the panzers' advance just short of Moscow by December 1941.

During the bitter defensive and offensive battles waged throughout 1942 and early 1943, the effective T-34 medium tank became the mainstay of the Red Army's tank force. Produced in the tens of thousands, the T-34 tank's effective balance of mobility, sloped-armoured protection, firepower, reliability, upgradability and economy of manufacture enabled this design to exert a significant impact on the Soviet Union's survival during the middle phase of this bitter struggle.

Alongside the iconic T-34s fought T-70 light tanks, KV-1E, KV-1 Model 1941/1942 and KV-1S heavy tanks, SU-122 SPGs and SU-76 assault guns; Western Allied-supplied Matilda, Valentine, Churchill and Stuart tanks augmented these assets.

Tank crews sit on their IS-2 heavy tank as units of the 1st Belorussian Front advance through East Prussia in early 1945. Abandoned German vehicles sit by the roadside.

Heavier tanks and SPGs

From late summer 1943 onward, moreover, the Soviet tank force – having wrested the strategic initiative from the Germans – began the protracted endeavour of driving the hated Axis invaders west out of Soviet territory.

New, more powerfully-gunned and up-armoured tanks and SPGs, produced in large numbers, contributed significantly to the final defeat of the Axis powers in May 1945. Many thousands of potent up-gunned T-34/85 medium tanks, plus some T-44s, played a key role, as did the KV-85 and IS heavy tanks. Backing up these forces were American-supplied Shermans as well as a range of powerful Soviet assault guns such as the SU-76M, SU-85, SU-100, SU-122 and SU-152, along with the ISU-122 and ISU-152.

By the war's end the Red Army tank corps had been transformed into the world's most powerful and effective armoured force, which operated some of the most powerful tank designs developed at that time.

This book describes these vehicles, exploring their firepower, levels of protection and mobility, while examining both their development and production histories.

ISU-122 self-propelled guns of the 1st Ukrainian Front sit camouflaged in a wooded area on the outskirts of Berlin, April 1945.

LIGHT TANKS

Of the total strength of 18,124 tanks that defended the Soviet Union in June 1941 against the onslaught of the German-led Axis Invasion called Operation 'Barbarossa', some 10,300 were light tanks.

This chapter includes the following models:

- T-18 Light Tank
- T-26 Model 1931 Light Tank (Twin Turret)
- T-26 Light Tank (Single Turret)
- KhT Light Flame Tanks
- T-37 and T-38 Amphibious Tanks
- T-40 Amphibious Light Tank
- T-50 Light Tank
- T-60 Light Tank
- T-70 Light Tank

Red Army soldiers ride on the back of T-60 light tanks as they cross the Neva River near Leningrad, January 1943. The Red Army widely deployed tank riders as infantry support during offensive operations.

T-18 Light Tank

The T-18 was the first tank designed and built by the Soviets. Its design relied heavily on the French Renault FT tanks that were captured in 1920 during the Russian Civil War.

Soviet domestic tank design began during 1925–28 with the development of the T-18 light tank, also termed the MS-1 in some documentation. The T-18 was a modification of the few French Renault FT tanks that the Soviets had captured from the 'Whites' back in 1920 during the hard-fought Russian Civil War.

T-18M
Crew 2
Production 1941 (200 vehicles converted)
Weight 5.95 tonnes (6.56 tons)
Dimensions 4.38m (14ft 4in) x 1.76m (5ft 9in) x 2.1m (6ft 11in)
Engine None; engine removed
Road Speed n/a
Range n/a
Armament 1 x 45mm (1.77in) 20K Model 1938 gun; 1 x 6.5mm (0.26in) co-axial Fedorov double-barrelled machine-gun
Armour 8–16mm (0.31–0.63in)

The T-18's improvements included an independent vertical-spring suspension and a modified, more reliable transmission. Powered by a Mikulin-designed 26kW (35hp) engine, the tank could reach the modest maximum road speed of 17km/h (10mph). The armour protection consisted of 16mm (0.63in) hull plates and six curved 8mm (0.31in) turret plates.

Armament

The tank's armament was the Soviet-manufactured copy of the French 37mm (1.46in) Model 1928 gun, which delivered 10–12 rpm, but which possessed poor armour-piercing capabilities. The turret front also featured a ball-mounted 6.5mm (0.26in) Fedorov double-barrelled machine-gun. During 1928–30 the Soviets built 960 T-18s, which served in the Red Army's first tank units right up until the start of World War II.

Upgunned tank

In the second half of 1941, in the wake of the stunning success achieved by the invading Axis forces during Operation 'Barbarossa', the desperate Soviets up-gunned 200 of the T-18s then in use in training establishments to create the T-18M. They fitted the old light tank with the long-barrelled 45mm gun (1.77in) 20K gun then mounted in the T-26.

To make space for the gun's rounds in the cramped interior these T-18Ms had their engine compartments removed; they were thus employed as static gun emplacements. Very few examples survived the fighting of 1941.

T-18M light tank

The T-18 tank had a distinctive appearance, given both its unusual lozenge-shaped tracks and its tall, centralized superstructure. Pictured is the up-gunned T-18M, which had its engine removed to accommodate the larger weapon.

T-26 Model 1931 Light Tank (Twin Turret)

In 1931 the T-26 light tank, based on the Vickers Medium 'E', entered service and ultimately became the mainstay of the Soviet tank force. With 10,117 built it became the world's most numerous tank at that time.

The T-26 Model 1931 was a tall tank with a raised central hull superstructure on top of which were mounted two side-by-side cylindrical one-man turrets that featured a distinctive raised rear section. In most Model 1931 tanks both turrets mounted a 7.62mm (0.3in) DT Model 1929 machine-gun. A few standard Model 1931 tanks, however, instead mounted a low-velocity short-barrelled 37mm (1.46in) PS1 gun in the right-hand turret. The tank possessed modest protection with 6–15mm- (0.24–0.59in-) thick armoured plates. Powered by a 66kW (88hp) petrol engine, the tank achieved a maximum road speed of 35km/h (22mph).

In addition to these vehicles, every command tank variant (the T-26TU Model 1931) also mounted this 37mm (1.46in) PS1 gun in its right-hand turret, while the left-hand one retained the DT machine-gun. Apart from this PS1 cannon, the T-26TU could also be differentiated from its standard cousin because it featured a distinctive long horizontal hull-rail antenna for its wireless-communications set. The T26TU was otherwise identical to the standard Model 1931 tank.

T-26A Model 1931

As well as its dual turrets, the T-26 featured a 'wedge-shaped' track arrangement that was taller at the vehicle's front than at the rear.

T-26 Model 1931
Crew 3
Production 1931–33
Weight 8.2 tonnes (9.13 tons)
Dimensions 4.9m (16ft 1in) x 3.4m (11ft 2in) x 2.36m (7ft 9in)
Engine 66kW (88hp) GAZ-T26 4-cylinder petrol
Road Speed 35km/h (22mph)
Range 130km (81 miles)
Armament 1 x 7.62mm (0.3in) DT Model 1929 machine-gun in each turret
Armour 6–15mm (0.24–0.59in)

T-26TU Model 1931 Command Tank

Along with the PS1 gun, the T-26TU featured a distinctive rail antenna that ran horizontally along the length of the hull superstructure's top edge.

T-26 Light Tank (Single Turret)

The next version of the T-26 was the Model 1933, which introduced several key modifications. The most obvious difference was that in this design the twin turrets were replaced by a large, centrally-positioned single turret.

T-26 Model 1933

The Model 1933 featured a single large two-man cylindrical turret with a rear-positioned rectangular storage bin. This turret mounted a high-velocity 45mm (1.77in) L/46 20K Model 1932 (later Model 1934) cannon. This gun produced a high muzzle velocity of 820m/s (2700ft/s), enhancing the Model 1933's lethality. Some 122 rounds for the main gun were housed in the vehicle. The turret also mounted a standard DT machine-gun co-axially.

Some late Model 1933 tanks had a second machine-gun mounted on a large external anti-aircraft housing on the turret roof, while other vehicles featured a third ball-mounted DT machine-gun on the turret rear. The T-26RT Model 1933, the command variant, featured a distinctive horseshoe-shaped rail antenna mounted around the turret top edge. Some 3439 Model 1933s were built during 1933–35.

T-26 Model 1936

As Model 1933 production unfolded the Soviets introduced minor improvements that simplified production, thus creating the new Model 1936. This variant featured the improved 20K Model 1934 gun. The tank's anti-personnel lethality was enhanced by most vehicles mounting three DT machine-guns: the standard co-axial weapon, a second ball-mounted one on the turret rear and a third housed in an anti-aircraft mount on the rear turret roof. Model 1936 tanks also featured two nighttime searchlights atop the gun mantlet, while some vehicles also featured a wireless radio and steel-rimmed road-wheels. Welding began to be used on parts of the vehicle instead of riveting.

T-26 Model 1938

The next upgraded T-26 was the Model 1938, which featured a redesigned conical turret. While this had the same 15mm (0.59in) plates as previously the turret's side plates were sloped, thus enhancing the tank's survivability. But to avoid making the design any heavier, and thus overloading the GAZ-T26 engine, the Model 1938 retained the Model 1936's hull. This variant also featured an improved version of the 45mm (1.77in) gun, the Model 1938, along with enhanced optical equipment. Factory 174 at Leningrad produced 800 Model 1938 tanks.

MODEL 1933 MID-PRODUCTION CHANGES

As with many of their tanks, the Soviet factories producing the T-26 incorporated modifications to the Model 1933 during its production run without creating a new year-designated Model. Later Model T-26 Model 1933s, for example, featured a new enlarged fuel tank that held 290 litres (64 gal) of petrol rather than the previous 182 litre (40 gal) tank. This modification increased the tank's maximum by-road operational range to 220km (140 miles).

T-26 Model 1933

The large single turret with long-barrelled 45mm (0.77in) gun differentiated the Model 1933 from its predecessor.

T-26 M1933

Crew 3

Production 1933–35 (3439 vehicles)

Weight 9.4 tonnes (10.36 tons)

Dimensions 4.9m (16ft 1in) x 3.4m (11ft 2in) x 2.31m (7ft 6in)

Engine 67kW (90hp) GAZ-T26 4-cylinder petrol

Road Speed 35km/h (22mph)

Range 130km (81 miles)

Armament 1 x 45mm (1.77in) 20K Model 1932 (later Model 1934) L/46 gun; 1 x co-axial 7.62mm (0.3in) DT Model 1929 machine-gun

Armour 6–15mm (0.24–0.59in)

T-26RT Model 1936 Radio Tank

After 1933 most T-26 command tank variants were differentiated from the standard T-26 battle tank by the large horseshoe-shaped antenna around the turret roof.

T-26 M1936

Crew 3
Production 1936–38
Weight 9.65 tonnes (10.6 tons)
Dimensions 4.9m (16ft 1in) x 3.4m (11ft 2in) x 2.31m (7ft 6in)
Engine 67kW (90hp) GAZ-T26 4-cylinder petrol
Road Speed 35km/h (22mph)
Range 220km (140 miles)
Armament 45mm (1.77in) 20K Model 1934 L/46 gun; 3 x 7.62mm (0.3in) DT machine-gun, 1 co-axial, 1 turret rear, 1 turret roof
Armour 6–15mm (0.24–0.59in)

T-26RT Model 1938 Radio Tank

The obvious external difference between the Model 1938 and its predecessor was its redesigned conical turret.

T-26 M1938

Crew 3
Production 1938–39
Weight 9.8 tonnes (10.8 tons)
Dimensions 4.9m (16ft 1in) x 3.4m (11ft 2in) x 2.31m (7ft 6in)
Engine 71kW (95hp) GAZ-T26 4-cylinder petrol
Road Speed 35km/h (22mph)
Range 220km (140 miles)
Armament 1 x 45mm (1.77in) 20K Model 1938 gun; 1 or 2 x 7.62mm (0.3in) DT machine-guns, typically co-axial, sometimes 1 turret roof
Armour 6–15mm (0.24–0.59in)

T-26 Model 1938

This atypical Model 1938, with three-tone brown-green-khaki summer camouflage, belonged to the 18th Tank Division of the VII Mechanized Corps.

T-26 Model 1939

Experience garnered during the July–August 1941 Battle of Lake Khasan, the border clash with the Japanese Army, led the Soviets to enhance the T-26's armour protection, creating the Model 1939. The redesigned conical turret featured a cast or drop-forged front section with an armoured under-turret box, while some of its plates were increased to 20mm (0.79in). Typically the Model 1939 lacked the second and third machine-guns usually seen in the previous two variants. The Model 1939 also featured a redesigned hull, with a widened superstructure that had thicker, sloping sides to augment survivability; the increased space enabled both larger fuel tanks to be fitted and the stowage of additional 45mm (1.77in) rounds.

T-26E

During the 1940 Soviet–Finnish Winter War even the up-armoured T-26 Model 1939s proved vulnerable to enemy fire. However, this was not true of the 80 tanks built during 1940 that were fitted with additional 30–40mm (1.18–1.57in) bolted-on armour appliqué plates to make the T-26E ('Ekranami' indicated 'screened armour'); a few earlier T-26 models were also retro-fitted with these plates. During the opening stages of 'Barbarossa', 9600 T-26 tanks of all types faced the Axis onslaught. By then the design's modest armour protection, as well as general vehicle wear, rendered the T-26 (including the latest variants) extremely vulnerable to German tank, anti-tank and artillery fire. Thousands were knocked out, broke down or abandoned during the 1941 campaign, although the Soviets continued to use the few obsolescent survivors in combat right through until 1943.

T-26 Model 1939

The Model 1939 was distinguished from its predecessor by its redesigned turret and lack of second and third machine-guns.

T-26 M1939

Crew 3
Production 1939–41 (2400 vehicles)
Weight 10.25 tonnes (11.3 tons)
Dimensions 4.9m (16ft 1in) x 3.4m (11ft 2in) x 2.31m (7ft 6in)
Engine 71kW (95hp) GAZ-T26 4-cylinder petrol
Road Speed 34km/h (21mph)
Range 220km (140 miles)
Armament 1 x 45mm (1.77in) 20K Model 1938 gun; 1, 2 or 3 x 7.62mm (0.3in) DT machine-guns (co-axial, turret roof, turret-rear)
Armour 6–20mm (0.24–0.79in)

T-26E (screened)

The rare T-26E, produced during 1940, featured bolted-on appliqué armoured plates to augment its survivability. This model has been painted in winter camouflage.

T-26E (screened)

Crew 3
Production 1940 (89 vehicles)
Weight 12 tonnes (13.3 tons)
Dimensions 4.99m (16ft 4in) x 3.5m (11ft 6in) x 2.31m (7ft 6in)
Engine 71kW (95hp) GAZ-T26 4-cylinder petrol
Road Speed 32km/h (20mph)
Range 220km (140 miles)
Armament 1 x 45mm (1.77in) 20K Model 1938 gun; 1 x coaxial 7.62mm (0.3in) DT machine-gun
Armour 6–20mm (0.24–0.79in) plus 30–40mm (1.18–1.57in) bolted-on appliqué plates

KhT Light Flame Tanks

The reliable T-26 design formed the basis of three flame-throwing tank variants. Belching forth their fearsome jets of flame, these vehicles were invariably much-feared by enemy infantrymen.

Given that the T-26 became the work-horse of the 1930s tank corps, the Soviets unsurprisingly also produced specialized T-26 AFV variants such as the KhT-26, KhT-130 and KhT-133 flame-throwing tanks.

The KhT-26 (OT-26) was the flame-throwing variant of the double-turreted T-26 Model 1931. However, in the KhT-26 the left turret was removed to enable larger fuel tanks and compressed air bottles to be fitted inside the vehicle's hull. The long-barrelled flame device was mounted on the remaining (right-hand) turret instead of the 37mm (1.46in) PS1 gun, alongside a co-axial 7.62mm (0.3in) DT machine-gun. The flame-thrower could deliver 70 one-second bursts of flame to a range of 30m (33ft). The Soviets produced 550 KhT-26s organised in 52-vehicle chemical tank battalions.

Similarly, the KhT-130 (OT-130) was the flame-throwing variant of the single-turreted T-26 Model 1933. It mounted either a short- or (more rarely) a long-barrelled flame-throwing device and retained the co-axial machine-gun; many vehicles also mounted an additional machine-gun in the turret rear. The Soviets produced 401 KhT-130s.

Finally, the Soviets produced 269 KhT-133 (OT-133) tanks, the flame-thrower variant of the T-26 Model 1938. This featured a short-barrelled flame device and also lacked the co-axial machine-gun. Some KhT-133s possessed bolted-on appliqué armour and an additional turret rear-mounted machine-gun.

CHANGING CAPABILITY

Up to 1939 the Soviets designated these three vehicles as the KhT-26, -130 and -133. The designation KhT was an abbreviation for the term 'Khimicheskiy tank' ('Chemical tank'). This terminology reflected the fact that when first designed these flame-throwing tanks were also intended for area chemical contamination and smoke-screen generation. These other capabilities were soon discontinued and by 1939 the three variants had been redesignated OT-26, -130 and -133, with OT being the abbreviation of 'Ognemetniy tank' ('Flame-throwing tank').

KhT-130 (OT-130)

A late KhT-130 equipped with a short-barrelled flame device.

KhT-133 (OT-133)

The KhT-133 was based on the T-26 Model 1938 and included a short-barrelled flame-throwing device.

KhT-133

Crew 3

Production 1938–40

Weight 9.8 tonnes (10.8 tons)

Dimensions 4.9m (16ft 1in) x 3.4m (11ft 2in) x 2.31m (7ft 6in)

Engine 71kW (95hp) GAZ-T26 4-cylinder petrol

Road Speed 35km/h (22mph)

Range 220km (140 miles)

Armament 1 x short-barrelled flame device; 1 x 7.62mm (0.3in) DT machine-gun, turret rear

Armour 6–15mm (0.24–0.59in)

T-37 and T-38 Amphibious Tanks

During the 1930s, the Soviets produced more Amphibious light tanks than any other nation; most of these modestly-armed and lightly-armoured tanks succumbed during the 1941 Axis invasion of the Soviet Union.

During 1931–32 the Soviets produced several experimental prototypes based on a Vickers export design. Despite this, they struggled to develop an effective Amphibious light tank for scouting missions.

T-37A

The lessons learned finally enabled the design of the satisfactory T-37A, which entered mass production in late 1933. The tall T-37A had a distinctive flat and shallow hull superstructure that sloped slightly upwards from the front of the vehicle toward the back. The T-37A featured a track suspension system based on a large frontal drive-wheel, two pairs of paired medium-sized horizontally-sprung six-spoke bogie road-wheels, a large six-spoke rear idler wheel and two track-top return rollers per side; the rear idler was positioned lower than the front driving-wheel.

The tank featured a tall cylindrical turret atop the hull and centre-right that mounted a 7.62mm (0.3in) DT machine-gun. With armoured plates of between just 3–10mm (0.12–0.39in) thickness, the vehicle weighed a mere 3.2 tonnes (3.5 tons). Powered by a 30kW (40hp) GAZ-AA engine that delivered a power-to-weight ratio of 9.4 kW/tonne (11.4hp/ton), the tank could develop a maximum road speed of 34km/h (22mph). To enable it to cross water the tank incorporated several specialized features, such as hollow track float-covers, rear mud-guard 'water wings' and a retractable three-bladed propeller.

Between 1933 and 1936 Soviet factories manufactured some 2600 T-37As, including 135 examples of the T-37TU variant (see box T-37TU Command Variant). T-37As and T-37TUs served as recce vehicles in Soviet armoured and cavalry brigades right through until summer 1942.

T-37A (winter camouflage)

This T37A tank has been finished in a mottled winter disruptor camouflage scheme, with dappled whitewash applied over its standard olive green finish.

T-37A

Crew 2

Production 1933–36

Weight 3.2 tonnes (3.5 tons)

Dimensions 3.75m (12ft 4in) x 2.1m (6ft 11in) x 1.82m (6ft)

Engine 30kW (40hp) GAZ-AA petrol (gasoline)

Road Speed 35km/h (22mph)

Range 185km (115 miles)

Armament 1 x 7.62mm (0.3in) DT Model 1929 machine-gun

Armour 3–10mm (0.12–0.39in)

T-37A

This side view ably illustrates both the tank's distinctive track arrangement and its large cylindrical turret.

T-37TU Command Tank

Soviet T-37TU command vehicles were distinguishable from other T-37 tanks by their prominent hull-top rail antenna.

T-37TU COMMAND VARIANT

In each T-37A platoon there would be eight ordinary battle tanks whose crews would communicate using large flags. The final vehicle, however, was the platoon commander's T-37TU command tank variant. This was almost identical to the standard tank except that it carried a wireless radio. The distinctive antenna that serviced this wireless device took the form of a long lateral rail that extended around the top of the hull superstructure's upper edge.

The hundreds of T-37s employed during the 1940 Winter War against Finland suffered heavy casualties. During 'Barbarossa', moreover, hundreds of these lightly-protected and under-gunned T-37s quickly succumbed to the Axis onslaught, even though few were ever employed in their amphibious role. Thereafter the surviving T-37s were relegated to training duties and replaced on the front line by the T-40.

T-38

Designed during 1934–36, the T-38 amphibious light scout tank was a development of the T-37A. It was a marginally squatter vehicle than its predecessor, but visually both tanks looked quite similar. The T-38s large cylindrical turret was positioned atop the centre-left of the rectangular flat hull roof. This turret mounted a standard 7.62mm (0.3in) DT machine-gun. The T-38s light 3–9mm (0.12–0.35in) armour kept its weight to just 3.3 tonnes (3.6 tons), making it air-transportable; in the 1936 exercises such air-portability was extensively tested.

The T-38 had the same 30kW (40hp) GAZ-AA engine as the T-37A, giving it an operational range of 170km (109 miles) and a top road speed of 40km/h (25mph). During 1936–39 the Soviets produced 1500 T-38s before manufacturing capacity was switched to its successor, the T-40. During 1938–39 some 150 up-gunned tanks, designated the T-38RT, were produced with the long-barrelled 20mm (0.79in) automatic-fire ShVAK cannon.

T-38

Crew 2
Production 1936–39
Weight 3.3 tonnes (3.6 tons)
Dimensions 3.76m (12ft 4in) x 1.95m (6ft 5in) x 1.62m (5ft 4in)
Engine 30kW (40hp) GAZ-AA petrol (gasoline)
Road Speed 40km/h (25mph)
Range 170km (109 miles)
Armament 1 x 7.62mm (0.3in) DT Model 1929 machine-gun
Armour 3–9mm (0.12–0.35in)

T-38 amphibious tank

The T-38 resembled the T-37A except that it had a shallower hull superstructure and more widely-spaced bogie-wheel pairings.

T-40 Amphibious Light Tank

During the late 1930s the latest Soviet Amphibious light tank design, the T-40, incorporated enhanced lethality and slightly greater survivability; these modest improvements, however, could not prevent the design still being out-classed by 1941.

During 1938–40 the Soviets designed an improved amphibious scout tank, the T-40, to replace the out-classed T-37A and T-38, which lacked adequate firepower and lethality. The T-40 was a squat rectangular vehicle with a tall, sloped-sided rectangular hull superstructure. The track suspension consisted of a large front track driving sprocket-wheel, four large non-bogie road-wheels, a large rear idler-wheel and three top-track return rollers.

On top of the superstructure's centre-left was a shallow conical turret that mounted a 12.7mm (0.5in) DShK heavy machine gun on the right as well as a co-axial 7.62mm (0.3in) DT machine-gun to the left. Firing the B-30 armour-piercing round, the DShK could penetrate up to 16mm (0.6in) of vertical armour plate at 300m (985ft). The T-40's armour protection was light, with just 3–13mm (0.12–0.51in) plates, although some of these were sloped.

Powered by a 63kW (85hp) GAZ-202 engine and with 100-litre (22-gal) fuel tanks, the T-40 could obtain an operational range of 450km (280mph) and a top road speed of 45km/h (28mph). A rear screw-propeller, two rudders and the tracks enabled the waterproofed T-40 to cross large rivers and lakes.

To speed production, from 1940 the Soviets also produced a simplified up-armoured land-only version, the T-40S, which lacked amphibious features and a radio, but also had turret armour increased to 20mm (0.79in). During 1940–41 the Soviets produced 448 T-40s, including 150 T-40S tanks.

Dire shortages of medium tanks during the winter of 1941/42 compelled the Red Army to use even the thinly armoured T-40 Model 1940 in an infantry support role.

T-40

Crew 2
Production 1939–41
Weight 5.5 tonnes (6.1 tons)
Dimensions 4.43m (13ft 6in) x 2.51m (8ft 3in) x 2.12m (6ft 11in)
Engine 63 kW (85hp) GAZ-202 petrol
Road Speed 45km/h (28mph)
Range 450km (280 miles)
Armament 1 x 12.7mm (0.5in) DShK Model 1938 heavy machine-gun; 1 x 7.62mm (0.3in) DT co-axial machine-gun
Armour 3–13mm (0.12–0.51in)

T-40 (winter camouflage)

This side view of a T-40 in whitewashed winter camouflage highlights its high sloped-sided hull superstructure.

T-60 Light Tank

The next generation of light tanks sported somewhat enhanced survivability thanks to their modest yet well-sloped armour; they also had increased firepower and were more economical to produce. Despite this they were still semi-obsolescent by summer 1941.

During the late 1930s the Soviets concluded that their T-26 light tanks were insufficiently armoured and not mobile enough to perform effectively on the contemporary battlefield, while the machine-gun-equipped T-26 Model 1933 vehicles were also too under-gunned to do so.

Therefore, during 1938 design work commenced on prototype replacement vehicles that would mount a more potent 20mm (0.79in) gun as well as light yet well-sloped armour. By 1941 this design work culminated in the emergence of the T-50 and T-60. While the diesel-powered T-50 had its production run curtailed after just 69 vehicles, that year the T-60 successfully entered into mass production.

Sloped armour

Weighing 5.2 tonnes (6.1 tons), this two-man vehicle featured a track suspension arrangement of one drive, four road and one rear idler wheel plus three return rollers on each side. The tank also sported angular well-sloped – albeit light – armour on the shallow hull superstructure; the latter gave the tank a low silhouette. The hexagonal angular turret, offset to the left, mounted a long-barrelled TNSh 20mm (0.79in) L/82.4 automatic-fire cannon and a co-axial DT machine-gun.

Crucially, the Soviets found that the T-60 could be easily and cheaply manufactured, which was crucial given the catastrophic tank casualties suffered during the 'Barbarossa' campaign in 1941. In a 14-month production run the Soviets delivered 6292 T-60s, including 1500 Model 1942 variants, before manufacturing capacity was switched to the improved T-70.

T-60 (early production)

The actual battlefield cross-country speed and mobility of the rather under-powered T-60 rarely lived up to Soviet expectations.

T-60 (late production)

The Model 1942 variant featured solid, rather than spoked, road-wheels, as well as thicker frontal armour.

T-60
Crew 2
Production 1941–43
Weight 5.2 tonnes (6.1 tons)
Dimensions 4.29m (14ft 1in) x 2.46m (8ft 1in) x 1.89m (6ft 2in)
Engine 52kW (70hp) GAZ-202 6-cylinder petrol
Road Speed 45km/h (27mph)
Range 435km (270 miles)
Armament 1 x 20mm (0.79in) TNSh L/82.4 automatic-fire cannon; 1 x 7.62mm (0.3in) co-axial DT machine-gun
Armour 7–35mm (0.28–1.38in)

T-50 Light Tank

While a promising new light infantry tank design, the T-60s was too expensive and complicated to mass produce in the conditions following the 'Barbarossa' invasion.

Alongside the T-60, in 1939–40 the Soviets also designed another potential replacement for the increasingly obsolescent T-26 and BT tanks, designated the T-50. This infantry support light tank featured well-sloped armour that made it appear externally as a sized-down version of the T-34. At 13.7 tonnes (15.1 tons) the T-50 was the heaviest Soviet pre-war light tank. The vehicle mounted a 45mm (1.77in) Model 1938 gun together with a co-axial 7.62mm (0.303in) DT machine-gun.

The tank's unusual power-plant comprised a V-4 diesel engine. With its torsion bar suspension the T-50 could develop a maximum road speed of 60km/h (37mph).

The tank possessed 12–37mm- (0.47–1.46in-) thick all-welded armoured plates; the protection provided was augmented by the deflective capabilities of this armour's 40–57 degree of slope. Although a promising design, the advent of war in 1941 curtailed production after just 69 had been completed. When maximum numbers of new tanks were desperately needed the tank was too challenging and costly to produce.

T-50 light tank

Although the T-50 featured a hull superstructure similar to that of the T-34, its wheel arrangement took a more traditional form with six road-wheels and three return rollers per side.

T-50
Crew 4
Production 1940–41
Weight 13.7 tonnes (15.1 tons)
Dimensions 5.2m (17ft 7in) x 2.47m (8ft 1in) x 2.16m (7ft 1in)
Engine 220kW (300hp) V-4 inline-six diesel
Road Speed 60km/h (37mph)
Range 220km (140 miles)
Armament 1 x 45mm (1.77in) Model 1938 gun; 1 x 7.62mm (0.3in) co-axial DT machine-gun
Armour 12–37mm (0.47–1.46in)

T-70 Light Tank

After the shock of the summer 1941 'Barbarossa' Axis invasion, the Soviets rushed the T-70 into mass production. It was essentially an up-armoured, up-gunned and up-powered modification of the existing T60.

In response to the shock of 'Barbarossa' during October 1941, the Soviets began developing an improved version of the T-60, designated the T-70. Given the catastrophic tank losses inflicted by the Axis invaders, the Soviets were desperate to rush this improved design into mass production. To expedite this rapid transition, the Soviets retained many of the T-60's design features. The GAZ factory completed the first prototype in December 1941 and after hurried pre-production trials full-scale T-70 production commenced in March 1942. The T-70's exterior was broadly similar to that of the T-60. It was a relatively low-silhouetted tank with a well-sloped hull superstructure sat upon a slightly 'wedged' track arrangement.

One-man turret

Positioned on the centre-left of the superstructure was an angular well-sloped one-man shallow conical turret. The T-70 was crewed by just two: the driver and the commander (who acted as loader and gunner). The T-70 was powered by a GAZ-203 engine – essentially two 52kW (70hp) GAZ-202s mated together along the tank's

T-70M
Crew 2
Production 1941–43
Weight 12.67 tonnes (14 tons)
Dimensions 4.29m (14ft 1in) x 2.32m (7ft 7in) x 2.04m (6ft 8in)
Engine GAZ-203 = 2 x 52kW (70hp) GAZ-202 6-cylinder petrol
Road Speed 45km/h (28mph)
Range 354km (220 miles)
Armament 1 x 45mm (1.77in) L/46 20K Model 1938 gun; 1 x 7.62mm (0.3in) co-axial DT machine-gun
Armour 10–45mm (0.39–1.77in)

T-70M (winter camouflage)

In late 1942 this whitewashed T-70M first entered combat, its red painted slogans crediting its funding via the Tambov Farmers' voluntary money-raising scheme.

T-70M (front view)

Although the T-70's turret was offset to the left, the gun's positioning meant that it remained close to the vehicle's centre-line.

right-hand side. The GAZ-203 enabled the T-70 to achieve a maximum road speed of 45km/h (28mph) and a peak operational range of 354km (220 miles).

Track suspension

The track suspension arrangement consisted of a large front track driving wheel, five large non-bogie doubled-tyred road-wheels, a large rear idler wheel and three relatively large top-track return rollers. The idler wheel, moreover, was positioned lower than the front drive wheel, which gave the T-70's tracks a modest 'wedge' shape. The vehicle's suspension was based on a strengthened torsion bar system. The inclusion of a fifth road-wheel, absent in the T-60, was one of several features that differentiated the T-70 from its predecessor.

The central innovation in the new T-70, however, was its main armament. The T-70 mounted the potent rapid-fire long-barrelled 45mm (1.77in) 20K Model 1938 gun (as fitted to the T-26 Model 1938 and 1939) in the vehicle's one-man turret instead of the T-60's less lethal 20mm (0.79in) weapon. The 20K Model 1938 cannon could obtain an impressive theoretical rate-of-fire of 12rpm, but with the commander having to load, aim, fire and command it seems doubtful that this rate-of-fire was frequently reached in combat. The heavy demands placed on the commander also hampered effective co-ordination between tanks at the platoon level.

In addition to its main gun the T-70 also featured a standard co-axial 7.62mm (0.3in) DT machine-gun. In addition, the tank carried 90 rounds for its 45mm (1.77in) gun. The T-70 also sported superior levels of protection, with sloped armoured plates up to 45mm (1.77in) thick on the turret. Subsequently the improved T-70M incorporated much enhanced running gear, a modified turret and reduced main armament round stowage. Between March 1942 and October 1943 Soviet factories produced 8225 T-70 tanks.

A column of T-70 light tanks advance through a winter blizzard on the Eastern Front.

MEDIUM AND FAST TANKS

Alongside the 10,300 light tanks fielded by the Red Army on the eve of Operation 'Barbarossa', the Soviets also fielded around 3000 BT fast tanks as well as 20 operational T-24 and 441 T-28 medium tanks.

This chapter includes the following models:

- T-24 Medium Tank
- T-28 Medium Tank
- BT-2 Fast Tank
- BT-5 Model 1933
- BT-5TU Model 1934
- BT-7 Model 1935
- BT-7 Model 1937
- BT-7M (BT-8)
- BT-7A Artillery Tank

A T-28 tank lies abandoned in Lithuania during the first week of Operation 'Barbarossa'. This vehicle is probably from the 5th Armoured Division, III Mechanised Corps.

T-24 Medium Tank

Soviet medium tanks – such as their first design, the T-24 – were more heavily armed and armoured than the Light tanks, but were also heavier and thus slower and less mobile.

Designed in 1931 the T-24 multi-turreted vehicle was the Soviet Union's first domestically-conceived medium tank. The vehicle's riveted main cylindrical turret, positioned well-forward on the central axis, housed a 45mm (1.77in) 20K Model 1932 gun on the left of the frontal turret face; to the right there was a ball-mounted 7.62mm (0.3in) DT machine-gun. Located atop the rear of this main turret was the cylindrical secondary turret, which mounted a second DT machine-gun. The tank also mounted a third DT machine-gun in a ball-mount located low down on the hull superstructure's near-vertical frontal plate. Delays in producing the 45mm (1.77in) gun meant that some T-24s were finished lacking this main armament, which was retro-fitted as guns came off the production line.

T-24
Crew 5
Production 1931 (25 vehicles)
Weight 18.2 tonnes (20.1 tons)
Dimensions 6.5m (7ft 3in) x 3m (9ft 10in) x 2.84m (9ft 4in)
Engine 190kW (250hp) M-6 petrol
Road Speed 25km (16mph)
Range 140km (90 miles)
Armament 1 x 45mm (1.77in) Model 1932 gun; 3 x 7.62mm (0.3in) DT machine-gun (1 turret front, 1 secondary turret, 1 hull front)
Armour 8–20mm (0.31–0.79in)

Heavy light tank

The combination of having two turrets as well as thicker 8–20mm (0.31–0.79in) armoured plates made the T-24 weigh a considerable 18.2 tonnes (20.1 tons). This left the vehicle more than twice as heavy as the T-26 Light Tank. The T-24's unusual track/suspension arrangement consisted of four paired very small bogie road-wheels, large front-driver and rear-idler wheels and four small top-track return rollers positioned atop the bogie assemblies. Like some early Soviet tanks, the T-24 had a removable 0.69m (2ft 3in) long tail that helped it cross wide trenches.

Underpowered engine

Powered by a 190kW (250hp) M-6 petrol engine, the comparatively under-powered T-24 could only obtain a modest maximum road speed of 25km/h (16mph). Early feedback from the extensive trials undertaken, as well as from its initial service employment, however, revealed that the T-24's power-plant and running gear experienced serious problems with mechanical unreliability. Consequently KhPZ's factory in Kharkov, Ukraine, prematurely halted T-24 production in late 1931 after a mere 25 tanks had been completed.

T-24
The T-24's most visually striking external features were its unusual running gear and the auxiliary turret atop the main turret.

T-28 Medium Tank

The T-28 became the standard Soviet medium tank design of the latter part of the 1930s; as soon as the T-34 medium tank entered service in 1940, however, the T-28 was totally eclipsed.

During 1930–31, the Soviets digested the lessons to be learned from the experience of testing the abortive T-24 medium tank. Over the subsequent three years, Soviet designers developed the multi-turreted T-28 Model 1934 medium infantry-support tank. This unusually long vehicle was well protected for its time. Its armoured plates varied between 10 and 30mm (0.39 and 1.18in) in thickness. This heavy armoured protection contributed to the tank's considerable 28-tonne (30.86-ton) weight. In terms of firepower, the T-28 featured a large cylindrical main turret with a large rear stowage bin. This main turret was positioned centrally atop the rather tall hull superstructure. Initially this turret mounted the 45mm (0.77in) 20K Model 1932 gun and a co-axial 7.62mm (0.3in) DT Model 1927 machine-gun. Many later T-28 medium tanks produced after 1936 mounted the short-barrelled 76.2mm (3in) KT-28 howitzer instead of the 45mm (0.77in) gun.

In addition to this main turret, the T-28 tank also featured two small cylindrical auxiliary turrets. These were located forward of and below the main turret in front of the vehicle's hull front plate. These turrets had an angular three-sided frontal section married to a semi-circular rear section. Both of these turrets, which were restricted to a 220-degree traverse, each mounted a 7.62mm (0.3in) DT machine-gun. Many Model 1934 tanks also featured a fourth machine-gun mounted in the turret rear face.

T-28 Model 1938

The improved Model 1938 variant included the longer 76.2mm (3in) L-10 gun as well as a machine-gun in the turret rear. This tank fought on the Karelian Front in the summer of 1941 and is painted in a three-tone camouflage.

T-28 front view

This frontal view of the T-28 illustrates the positioning of the two auxiliary turrets below and in front of the main turret.

T-28 Model 1934

Crew 6
Production 1934–40
Weight 28 tonnes (30.86 tons)
Dimensions 7.44m (24ft 5in) x 2.81m (9ft 3in) x 2.82m (9ft 3in)
Engine 373kW (500hp) V-12 Milukin-M17 petrol
Road Speed 37km/h (23mph)
Range 220km (137 miles)
Armament 1 x 45mm (0.77in) 20K Model 1932 gun; later short-barrelled 76.2mm (3in) KT-28 howitzer; 3 x 7.62mm (0.3in) DT Model 1927 machine-guns
Armour 10–30mm (0.39–1.18in)

Despite this bulk, the tank's potent 373kW (500hp) V-12 Milukin-M17 45-litre (9.9 gal) petrol engine enabled the T-28 to reach a maximum cross-country speed of 37km/h (23mph). The T-28's track/suspension system comprised six pairs of small bogie-mounted rubberized road-wheels on plunger coil springs (which were partially hidden beneath 20mm/0.79in thick side-skirting armour plates), four top-track return rollers, a large front idler wheel and a medium-sized rear drive-wheel.

Subsequently, during 1938 the slightly modified T-28 Model 1938 entered production. This variant mounted the longer-barrelled 76.2mm (3in) L-10 gun as standard, which delivered improved armoured-piercing capabilities.

Finally, during 1939–40, amid the Winter War with Finland, modified T-28s (designated T-28E or -28M) appeared with additional bolt-on appliqué armour designed to augment their survivability. T-28 production continued at a slow pace for a seven-year period that stretched from 1934 right through to 1940. During this period the Soviets produced just 503 T-28 medium tanks of all variants.

SIDE-SKIRTS
The T-28 featured armoured side-skirts fitted to the side of the vehicle to protect its track and suspension system. In the T-28, the top line of the sloping top skirt row featured small indentations so that it fitted snuggly around the four return rollers; beneath this the tank sported either one or two rows of vertical armoured plate that extended right down to cover the top of its 12 pairs of road-wheels.

BT Fast Tanks

A prevalent feature of 1930s Soviet tank doctrine was the significance placed on the fast tank concept. Such vehicles relied on their impressive speed and mobility to survive, while conducting rapid exploitation-style operations.

BT-2

Soviet armoured warfare doctrine emphasized mobility to rapidly exploit success. The Soviet Light tanks of the 1930s partially fulfilled this role, but they lacked sufficient firepower, protection and mobility. Therefore, the Soviets designed the BT fast tank series. These were marginally heavier and better armed than the Light tanks, but were also much faster, which boosted survivability.

The BT tanks' impressive speed stemmed from their use of the radical American-designed Christie suspension system. This featured vertical-coil springs that supported the large non-bogie road-wheels around which the track ran tightly, making upper track return rollers unnecessary.

Based on the M-1930 Christie chassis, the three-man BT-2 fast tank entered production in 1932. It was

BT-2
Crew 3
Production 1932–34
Weight 10.2 tonnes (11.24 tons)
Dimensions 5.58m (18ft 4in) x 2.3m (7ft 7in) x 2.25m (7ft 5in)
Engine 298kW (400hp) M5 petrol
Road Speed 100km (62mph)
Range 200km (124 miles)
Armament 1 x 37mm (1.46in) Model 1930 gun; 1 x 7.62mm (0.3in) DT machine-gun
Armour 6–13mm (0.24–0.51in)

BT-2
This side view illustrates the four large open-spoked road-wheels per side featured in the BT-2's Christie-style suspension. This tank has been daubed with white paint as a rudimentary winter camouflage.

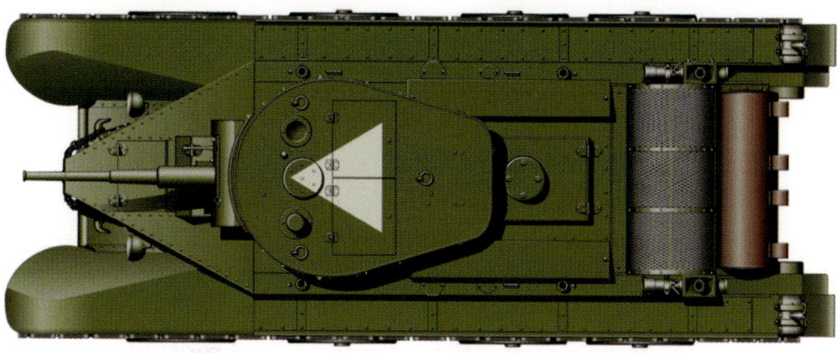

BT-5 Model 1933

The BT-5 utilized the same chassis as the BT-2 but mounted a redesigned curved-sided turret often fitted with a rear stowage bin.

BT-5 Model 1933
Crew 3
Production 1933–35
Weight 11.5 tonnes (12.7 tons)
Dimensions 5.58m (18ft 4in) x 2.3m (7ft 7in) x 2.25m (7ft 5in)
Engine 298kW (400hp) M-5 petrol
Road Speed 72km/h (45mph)
Range 200km (124 miles)
Armament 45mm (1.77in) 20K Model 1932 gun; 1 x 7.62mm (0.3in) co-axial DT machine-gun
Armour 6–13mm (0.24–0.51in)

---

a long, low-silhouetted tank with a sloping wedge-shaped glacis plate and a conical turret situated well forward on the central axis. Most BT-2s mounted the 37mm (1.46in) Model 1930 gun with a DT machine-gun in a ball-mounting on the turret right front. Some BT-2s, however, had a double DP machine-gun mounting in place of the main gun while retaining the ball-mounted machine-gun. Weighing 10.2 tonnes (11.2 tons), the BT-2 had modest armoured plates of 6–13mm (0.24–0.51in) thickness. The tank's survivability was enhanced, however, by its astonishing 100km/h (62mph) maximum road speed, delivered by a powerful 298kW (400hp) Milukin M-5 petrol engine.

The tank could operate cross-country using tracks or along roads just using its wheels; the crew could switch configurations in 30 minutes. Plagued by poor mechanical reliability, however, BT-2 production halted in 1934 after 620 vehicles were made.

BT-5

The next fast tank design, the BT-5 Model 1933, entered service that same year. This three-man tank featured a redesigned turret on top of a hull identical to that of the BT-2. The new curved-sided cylindrical turret was positioned towards the front of the tank on the central axis. Many later BT-5 turrets also had a large and long stowage bin attached to the back of the turret, as well as welded rather than riveted construction. This new two-man turret mounted a new, more powerful gun, the 45mm (1.77in) Model 1932, as well as a co-axial DT machine-gun; the new cannon provided the tank with a much-increased anti-armour capability.

The BT-5 Model 1933 weighed 11.5 tonnes (12.7 tons) and possessed the same modest armour as the BT-2, with 6–13mm (0.24–0.51in) armoured plates, barely adequate to withstand even armour-piercing rifle rounds. Despite the new turret, the vehicle's dimensions remained the same as its predecessor, the BT-2.

The tank was powered by the same powerful Mikulin-designed 298kW (400hp) M5 petrol engine. However, with the new turret adding 1.3 tonnes to the vehicle's weight, and with smaller fuel tanks, the vehicle's top road speed was restricted to 72km/h

'CONVERTIBLE' TANKS

The BT fast tanks were 'convertible' vehicles: they could move both when using their tracks and with the tracks removed. Based on the Christie design, the crews of BT tanks could swiftly remove the tracks (which were then stowed) and the tank could run on just its wheels when travelling along good-quality roads. By 1940 the Soviets had concluded this this was an expensive, weight-consuming and little-used 'luxury' that was discontinued in the T-34 and subsequent tanks.

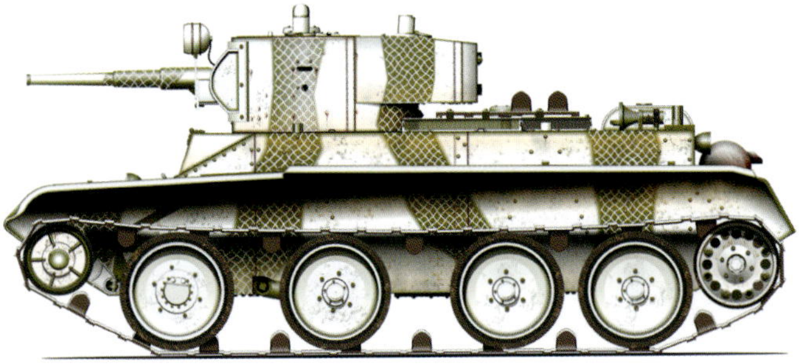

BT-5 Model 1934 (winter camouflage)

This BT-5 has been finished in two-tone camouflage, designed for the Moscow campaign of winter 1941–42.

BT-5 Model 1934

The BT-5 had a powerful engine, the Soviet-designed M-5 12-cylinder aircraft model, specially adapted for tank use.

(45mph) with a 200km (124 mile) operational range; despite this the Soviets still principally relied on the BT-5's fast speed for survivability. The BT-5 also featured redesigned large convex road-wheels. During 1933–1935 the Soviets produced 1884 Model 1933s.

BT-5TU Model 1934

As per usual Soviet practice, the KhPZ factory in Kharkov devoted a proportion of its BT-5 production capacity to deliver the command tank variant, the BT-5TU Model 1934. These tanks were in most respects identical to the standard BT-5 combat tank. However, one of the key differences was that the BT-5TU carried a two-way wireless radio communications set, the 71-TK-1, unlike the standard BT-5 tank. The antenna that served this radio device was not the the archaic hull-top arrangement seen on the early T-26TU Model 1931 command tank design. Instead, the BT-5TU featured the same large horseshoe-shaped antenna running around the top edge of the turret as fitted to the T-26RT Model 1933 command tank; in the BT-5TU the antenna rail was attached to the turret sides by three short struts.The size of this antenna arrangement was considerable – the width of the horseshoe was almost as wide as the entire width of the tank, while its length was greater than that of the turret. The 71-TK-1 radio system comprised two rectangular sets, one the transmitter and the other the receiver. The set had a maximum range of about 8km (5 miles) in

BT-5TU Model 1934 Command Tank

Crew 3

Production 1934–35 (80 vehicles)

Weight 11.55 tonnes (12.73 tons)

Dimensions 5.58m (18ft 4in) x 2.3m (7ft 7in) x 2.25m (7ft 5in)

Engine 298kW (400hp) M-5 petrol

Road Speed 70km/h (44mph)

Range 190km (118 miles)

Armament 45mm (1.77in) 20K Model 1932 gun; 1 x 7.62mm (0.3in) co-axial DT machine-gun

Armour 6–13mm (0.24–0.51in)

BT-5TU Command Tank

Finished in a two-tone olive and off-white camouflage scheme, this vehicle is the command variant of the BT-5 with its distinctive rail antenna.

Tank repairs in the depth of winter posed major problems for the Red Army. Here the turret is lifted from a BT-5 in a field workshop during the winter of 1941/42.

optimal climatic conditions, although in poor weather the effective range was probably about half this. During 1934–36 KhPZ manufactured 80 BT-5TUs, sufficient to equip most company commander's vehicles, as well as a tiny proportion of the tanks used by platoon commanders.

BT-7 Model 1937

Exploiting the experience gained with the BT-2 and BT-5, during 1935–37 the Soviets produced 150 pre-production vehicles for a new fast tank design, the BT-7 Model 1935. This tank mounted the existing BT-5 turret on a modified and enlarged BT-5 chassis. A few of these vehicles were produced as commander's tanks and designated the BT-7RT (or BT-7TU according to some sources) Model 1935 tank. They were distinguished by the large horseshoe frame antenna around the turret top.

After further modification this design entered mass production at KhPZ designated the BT-7 Model 1937. This combined the redesigned BT-7 Model 1935 chassis with a new turret. The Model 1937 tank featured a new conical turret design with 15mm- (0.59in-) thick sloped sides, derived from that on the T-26 Model 1938. The turret mounted a 45mm (1.77in) 20K Model 1932 gun as well as a co-axial 7.62mm (0.3in) DT machine-gun. The new chassis was slightly larger than that of the BT-5, with its length increased to 5.66m (18ft 7in) and its width to 2.29m (7ft 6in).

These modifications aside, the BT-7 Model 1937 featured the same track and suspension system as the BT-5. The slightly enlarged chassis and new turret increased the tank's weight to 14 tonnes (15.4 tons). However, the installation of a more powerful engine, the 373kW (500hp) Milukin M-17T, increased the design's range

BT-7RT Model 1935

The horseshoe-antenna rail, BT-5 turret and modified chassis identifies this tank as the BT-7RT Model 1935.

BT-7 Model 1937

This, the mass-production version of the BT-7, featured a redesigned sloping-sided conical turret. This tank has been painted with three-tone camouflage and served in eastern Poland in 1939.

BT-7 Model 1937
Crew 3
Production 1937–39
Weight 14 tonnes (15.4 tons)
Dimensions 5.66m (18ft 6in) x 2.29m (7ft 6in) x 2.42m (7ft 11in)
Engine 373kW (500hp) Milukin M17-T V-12 petrol
Road Speed 86km/h (53mph)
Range 250km (155 miles)
Armament 45mm (1.77in) 20K Model 1932 gun; 1 x 7.62mm (0.3in) co-axial DT machine-gun
Armour 6–15mm (0.24–0.59in)

to 250km (155 miles). Late-production vehicles (termed the Models 1938 and 1939 in some sources) featured minor improvements including the TOS gun-sighting system, reinforced hull braces and a hull floor escape hatch. In total, during 1937–1939 KhPZ produced 4900 BT-7 Model 1937 tanks.

BT-7M (BT-8)

During 1938 the KhPZ factory produced four experimental upgraded BT-7s that were designated the BT-8. These were essentially BT-7 Model 1937 tanks but, unlike most existing Soviet tanks, they were powered by a diesel engine rather than a petrol one. The engine incorporated in the BT-8 design was a 336kW (450hp) liquid-cooled V-12 diesel one based on an Hispano-Suiza 12Y aircraft design; the power plant was built separately from the tank at the Voroshilovets factory in Vladivostok.

These four vehicles were subjected to lengthy trials, which soon established that the potent new diesel power plant significantly increased the BT-8s maximum operational range to an astonishing 708km (440 miles).

Consequently, the Soviets decided to replace existing BT-7 Model 1937 production with the BT-8 design, albeit with the inclusion of some further modifications. These included the mounting of the improved 45mm (1.77in) 20K Model 1938 gun as well as increased armoured protection. Model 1937 tanks also featured an additional machine-gun in the turret rear and often a further one held in an anti-aircraft mount on the turret roof. When the BT-8 entered mass production in late 1939 it was re-designated the BT-7M (or the BT-7 Model 1940 according to some sources). The BT-7M also featured a slightly redesigned hull, with a modified full-length sloping glacis plate being introduced.

Between 1939 and mid-1941 Soviet factories produced a total of 790 BT-7M tanks. By the start of Operation 'Barbarossa' in June 1941, there were around 2500 BT-7s in service, mainly Model 1937s and BT-7Ms; over 80 per cent of these tanks were destroyed,

BT-7M (BT-8)

The sloping-sided conical turret and turret-rear machine-gun identify this as a BT-8, the diesel-powered, final variant of the BT-7. This tank served on the Leningrad Front during 1943.

BT-7M (BT-8)
Crew 3
Production 1939–41
Weight 14.7 tonnes (16.2 tons)
Dimensions 5.66m (18ft 6in) x 2.29m (7ft 6in) x 2.42m (7ft 11in)
Engine 336kW (450hp) liquid-cooled V-12 diesel
Road Speed 86km/h (53mph)
Range 708km (440 miles)
Armament 45mm (1.77in) 20K Model 1938 gun; 2 or 3 x 7.62mm (0.3in) DT machine-gun, (1 co-axial, 1 turret rear, 1 turret roof)
Armour 6–22mm (0.24–0.86in)

BT-7A Artillery Tank
Crew 3
Production 1936–38
Weight 14.5 tonnes (16 tons)
Dimensions 5.66m (18ft 6in) x 2.29m (7ft 6in) x 2.52m (8ft 3in)
Engine 373kW (500hp) Milukin M17-T V-12 petrol
Road Speed 86km/h (53mph)
Range 250km (155 miles)
Armament 76.2mm (3in) KT-28 Model 1932 howitzer; 1 or 2 7.62mm (0.3in) DT machine-gun (1 turret roof, 1 turret rear)
Armour 6–15mm (0.24–0.59in)

BT-7A Artillery Tank
To house the 76.2mm (3in) howitzer mounted in the BT-7A assault tank, this vehicle needed a redesigned and enlarged drum-shaped turret.

captured or abandoned during the bitter battles that raged during 1941.

BT-7A

As per normal practice, Soviet factories produced a number of BT-7 variants other than the BT-7TU command tank. One of the more important of these variants was the BT-7A self-propelled howitzer; the 'A' suffix indicated artillery tank in Russian parlance.

The BT-7A featured a redesigned and enlarged drum-shaped turret in comparison with the standard BT-7 tank, which was based on that mounted on the T-28. This new turret was mounted upon a hull identical to that of the BT-7. Additional internal turret space was required to fit a larger weapon, the short-barrelled 76.2mm (3in) KT-28 L/16 howitzer. The tank carried 50 rounds for its main armament. The turret top also mounted a machine-gun on an anti-aircraft mount; unusually for Soviet AFVs in some BT-7As this was positioned toward the front rather than rear. In other BT-7A tanks a ball-mounted machine-gun in the turret rear was also fitted in addition to the anti-aircraft machine-gun.

The KhPZ factory, moreover, produced about one-tenth of the BT-7A production run as platoon commander's vehicles. These carried a wireless radio set, with the distinctive horseshoe-shaped rail antenna along the tank's turret top edge. Because of the large space required to house the radio set, these vehicles only carried 40 rounds for its 76.2mm (3in) gun.

Between 1936 and 1938 the KhPZ factory at Kharkov produced 154 BT-7A artillery tanks, including around 15 commander's tanks; a number were subsequently up-graded to the specifications of the BT-7M.

Anti-aircraft mount
The BT-7A included a 7.62mm (0.3in) DT machine gun anti-aircraft mount.

T-34 MEDIUM TANKS

During October 1941 T-34 medium tank attacks temporarily halted the invading German panzers. With its excellent trade-off between lethality, survivability, mobility, usability and ease of production, the T-34 became the mainstay of the Soviet tank force, with 57,300 vehicles manufactured during 1940–45.

This chapter includes the following models:

- A-20 prototype
- T-34 Model 1940
- T-34 Model 1941
- T-34 Model 1942
- T-34 Model 1943
- T-34/85 Medium Tank
- OT-34 Flame Tank
- T-44 Medium Tank

A column of whitewashed T-34/85 tanks wait by the roadside during the winter of 1944–45. The wide tracks of the T-34 made them ideal for operations in the snow and mud of a Russian winter.

A-20 prototype

During the late 1930s Soviet tank design, such as the A-20, leapt forward by combining four elements: diesel engines, well-sloped armour, more lethal main armament and the abandonment of the weight-consuming non-track road movement capability.

Soviet participation in the 1936–39 Spanish Civil War revealed both that their existing tanks were vulnerable to the latest anti-tank guns, whose hits frequently caused the tank's petrol engines to catch fire. Consequently, during 1937 the Soviets began designing an innovative new medium tank design – the A-20 – with greater survivability and firepower than existing tanks. The 20.2 tonne (22.3 ton) A-20 incorporated novel well-sloped 10-20mm- (0.39-0.79in-) thick armoured plates, which enhanced its survivability. The A-20 also mounted the 45mm (1.77in) 20K Model 1938 gun then used in the existing BT-8.

The A-20 was powered not by a standard petrol engine but instead with the new 373kW (500hp) Model V-2-34 diesel engine, which enabled the tank to obtain a maximum by-road speed of 85km/h (53mph). The prototype also featured a Christie suspension with four large road-wheels and no return rollers per side; it retained the capability to travel on roads without tracks. The A-20 performed well during trials carried out in early 1938.

A-20

The A-20 prototype had a Christie suspension that supported four road-wheels, well-sloped armour and a diesel engine.

A-20
Crew 4
Production 1937–38
Weight 20.2 tonnes (22.3 tons)
Dimensions 5.7m (18ft 8in) x 2.7m (8ft 10in) x 2.4m (7ft 10in)
Engine 373kW (500hp) Model V-2 diesel
Road Speed 65km/h (40mph)
Range 350km (217 miles)
Armament 1 x 45mm (1.77in) 20K Model 1938 gun; 2 x 7.62mm (0.3in) DT machine-guns
Armour 10–20mm (0.39–0.79in)

T-34 Model 1940

Building upon the experience gained with prototypes such as the A-20, by 1940 the Soviets had introduced the first variant of what would become the mainstay of their tank force – the famous T-34 medium tank.

T-34 Model 1940 (cast turret)

The Model 1940 tank mounted two slightly different turret designs; this is the cast version.

Although the Soviets were satisfied with the A-20's performance, the July–August 1938 Soviet–Japanese clash at Lake Khasan convinced the Russians that even their latest tanks lacked adequate lethality and

Right: A column of T-34 Model 1941 tanks assemble somewhere in southern Ukraine, late summer 1941.

survivability. Consequently, the Soviets designed two heavier up-gunned and up-armoured versions of the A-20, designated the A-32 and A-34. These designs abandoned capability to travel on roads using just wheels not tracks. This capability added unnecessary weight to the tank since it was seldom used in reality. These vehicles mounted a more potent longer-barrelled 76.2mm (3in) gun and well-sloped 30–60mm (1.2–2.4in) armoured plates.

A-34 model

During June 1940 the four-man A-34, now designated the T-34 Model 1940 (T-34A), entered mass production. The T-34 reflected an impressive balance between the requirements of firepower, protection, mobility, good ergonomics, reliability and ease of production. Visually, the 26 tonne (28.7 ton) tank had well-sloped sides to its rectangular hull, an extended sloped frontal glacis plate and a sloped rear hull plate. The hexagonal-shaped welded or cast two-man turret, which was mounted well-forward on the superstructure, also had well-sloped sides. The slope enhanced the survivability provided by the tank's 15–45mm (0.59–1.77in) armoured plates.

On the turret roof there was a single large forward-opening hatch, which proved unpopular with tank commanders because it constrained their field of vision. In the turret's well-rounded front nose the tank mounted the 76.2mm (3in) L-11 L/30.5 cannon in a cylindrical box mantlet. This relatively short-barrelled weapon only generated a fairly modest armour-piercing capability. The turret also mounted a co-axial 7.62mm (0.3in) DT machine-gun, while a second such weapon was mounted in the hull front.

Some Model 1934s featured the 373kW (500hp) V-2-34 diesel engine, although owing to shortages other vehicles instead had the BT-7's MT-17 petrol engine. The T-34 also featured an improved coil-spring Christie suspension married to a track arrangement that comprised one small front driver-wheel, five large ten-spoked open metal road-wheels, one medium-sized rear idler positioned at the same level as the road-wheels and no return rollers. This enabled the tank to reach a maximum road speed of 55km/h (34mph) and a by-road operational range of 300km (186 miles).

Around one-tenth of the 400 Model 1940 tanks produced were commander's vehicles equipped with the 9-RS wireless radio set serviced by a whip antenna.

T-34 Model 1940 (welded turret)

The slight differences evident in this Model 1940 tank's turret bottom sides indicate that it sports the welded design.

T-34 Model 1940
Crew 4
Production 1940–41
Weight 26 tonnes (28.7 tons)
Dimensions 5.92m (19ft 5in) x 3m (9ft 10in) x 2.45m (8ft)
Engine 373kW (500hp) Model V-2-34 diesel
Road Speed 55km/h (34mph)
Range 300km (186 miles)
Armament 1 x 76.2mm (3in) L-11 L/30.5; 2 x 7.62mm (0.3in) DT machine-gun, 1 co-axial, 1 hull front
Armour 15–45mm (0.59–1.77in)

but incorporated minor design modifications to simplify production. Most Model 1942s sported a redesigned turret. Many UZTM-manufactured T-34Cs featured a 'stamped' single-piece pressed hexagonal two-man turret made from 45mm (1.77in) hardened steel. These turrets were distinguished by their rounded edges, the slight downward curve of the turret's central bottom edge and the gun mantlet's vertical bottom face. 'Stamped' pressed turret production speeded manufacture and improved survivability. Both the pressed and cast UZTM-manufactured turrets were also sent to ChTZ and Factory #112 for assembly onto other T-34 chassis. In contrast, STZ-produced T-34s had the traditional hexagonal cast or welded two-man turret. During the summer of 1942 STZ mainly produced Model 1941/42 tanks, incorporating a few of the Model 1942's features. STZ turrets featured a box-shaped gun mantlet distinguished by its outwards-sloping flat-faced design.

New hull design

The Model 1942 was also differentiated from its predecessors due to its redesigned sloping hull front glacis plate. The new driver's hatch, located on the left, had two armoured periscopes, not one. On the right, a new armoured gun-shield protected the hull machine-gun.

Because the factories altered some Model 1942 design features during their respective production runs, a wide variety of modifications soon emerged. Many Model 1942s were fitted with all-metal road-wheels, yet some had rubber-rimmed ones. Some tanks had 500mm (19.7in) wide tracks, while others were fitted with 550mm (21.7in) tracks. Some T-34Cs had rectangular box-shaped external fuel tanks fitted to the rear hull sides, a few had cylindrical ones and others had none. A few late-production Model 1942s also featured twin turret roof hatches. These design changes and variations increased the

T-34 Model 1942

Built in the Stalingrad Tractor Zavod (STZ), this tank dates from the early summer of 1942. It includes both older design elements (all-steel-road wheels, rectangular engine hatch, front machine gun mount), and newer features (engine deck side louvres, front and rear hooks). The four views also illustrate the sloped armoured plates that were key to the design's survivability.

UZTM'S 'SHLEMAN' PRESS

The T-34 Model 1942 tanks manufactured at UZTM's 'Uralmash' factory in Sverdlovsk had distinctive turrets with rounded edges, a slightly waved bottom edge and a non-flat-faced gun mantlet. This was thanks to the presence of 'Shleman,' the huge 9071 tonne (10,000 ton) steel press employed on the UZTM plant's production line. This press could transform a single large piece of 45mm (1.77in) thick hardened steel into a 'stamped' pressed turret.

tank's weight to 28.5 tonnes (31.4 tons). In total, some 11,000 Model 1942s were manufactured.

PT-34 mine-rolling tank

One of the principal specialized variants of the 76mm (3in)-gunned T-34 was the PT-34 Engineer Tank. In May 1942 the first experimental mine-clearing detachments were formed and equipped with newly-converted PT-34s.

This specialized design was a standard T-34 Model 1942 tank fitted with a rotating mine-clearing 'trawl' device. Two Mugalev-designed jib-booms projected forward and slightly upwards from the central portion of the standard T-34 tank's lower hull front. At the devices' front-end the jib-booms joined one another to form a housing; through this ran a horizontal axle that mounted on each side five rotating large and heavy circular metal 'serrated' discs. The 'serrations' were H-shaped metal forms that rotated out from the disk and dug into the ground as the disk rotated. As the tank advanced, these ten disks rotated along the ground, detonating any anti-tank mines planted by the enemy.

From October 1943 these mine-clearing detachments were consolidated into 40-vehicle engineer tank regiments; these fielded 18 PT-34 tanks supported by 22 standard T-34 tanks. The PT-34 tank retained its functional main armament to defend itself, although tactical manoeuvre was difficult with the mine-clearing device attached.

From late 1943 this mine-rolling device was mounted on T-34/85 tanks, with these vehicles also being designated PT-34.

PT-34 Engineer Tank

This side view of a PT-34 illustrates the large size of the mine-clearing 'trawl' device carried in front of the vehicle.

T-34 Model 1943

The last T-34 variant equipped with the 76mm (3in) gun was most easily distinguished from its predecessors by the addition of a large drum-shaped commander's cupola on the turret roof.

In early 1943, further sequential modifications to the existing T-34 design were incorporated into the production process, resulting in vehicles retrospectively designated the Model 1943. This variant was known to the Germans as the T-34D, T-34E and T-34F tank.

These modifications only slightly altered the tank's specifications; instead, they simplified and speeded-up the production process as well as improved the vehicle's combat usability. The T-34 Model 1942's key features – the 76.2mm (3in) F-34 gun, the two machine-guns, the 20–70mm (0.79–2.76in) sloped-armour, the 373kW (500hp) diesel engine and the five road-wheel Christie track/suspension arrangement – remained unaltered.

However, the modifications and alterations did raise the Model 1943's weight to 30.9 tonnes (34.1

tons). Although its top by-road speed remained 55km/h (34mph), the inclusion of larger external fuel tanks increased its maximum by-road operational range to 465km (289 miles).

New turret cupola

The first Model 1943 tanks (T-34D) had the same 'Mickey Mouse' turret-roof twin hatches as seen in the final Model 1942 production batch. However, the main externally obvious new feature in most subsequent Model 1943s (also known as T-34-Es) was located on the vehicle's turret roof. Here a raised drum-shaped commander's cupola was included on the left of the turret roof to replace the previous flush-level commander's hatch. This innovation afforded the vehicle commander better all-round observation and thus improved his situational awareness and tactical capabilities. The inclusion of improved main armament optics also increased the Model 1943 tank's tactical efficiency.

Whereas many Model 1942 tanks had all-steel road-wheels, most Model 1943 tanks featured a particular mix of

T-34 Model 1943 (summer camouflage)

An early T-34 Model 1943 with a round-edged turret that lacks a drum cupola on the turret roof is finished off in a striking four-tone 'disrupter' camouflage pattern.

T-34 Model 1943 (winter camouflage)

This early Model 1943 is finished in a mottled 'winter disruptor' camouflage scheme that combines olive with whitewash.

T-34 Model 1943
Crew 4
Production 1943–44
Weight 30.9 tonnes (34.1 tons)
Dimensions 5.92m (19ft 5in) x 3m (9ft 10in) x 2.45m (8ft)
Engine 373kW (500hp) Model V-2-34 diesel
Road Speed 55km/h (34mph)
Range 465km (289 miles)
Armament 1 x 76.2mm (3in) F-34 L/40.5 gun; 2 x 7.62mm (0.3in) DT machine-gun; 1 co-axial, 1 hull front
Armour 20–70mm (0.79–2.76in)

Left: A column of T-34 Model 1943 tanks passes through a Ukrainian village, summer 1943.

T-34 Model 1943

A more 'typical' Model 1943 that includes the distinctive commander's drum cupola and external fuel tanks; it also features open-spoked road-wheels instead of full-disk ones. This tank sports the name of an Armenian folk hero, 'David of Sasun', on the turret. The funds to build all the tanks for the 119th Tank Regiment were contributed to by donations from Armenian emigrants.

road-wheel designs: the first and last (fifth) roads-wheels had rubberized rims, whereas the remaining three were all made entirely from steel. Despite chronic shortages of rubber, the experience of the Model 1942 with five all-steel wheels had shown that vibration problems occurred with this arrangement. Experiments showed that having rubber-edged first and last wheels much reduced this problem.

Turret styles

As with the previous model, Model 1943s continued to feature three slightly different hexagonal two-man turrets: the UTZM pressed 'stamped' turret with rounded edges, the hard-edged cast and the soft-edged cast turret. The Germans designated T-34 tanks with the UTZM pressed turret as the T-34F. Increasingly larger proportions of the Model 1943 production were outfitted with a 9R wireless radio set, supported by a whip antenna mounted on the front-right hull edge. Most Model 1943s also incorporated long cylindrical external fuel tanks mounted on the rear hull sides instead of the rectangular box-like tanks mounted on the Model 1942.

During 1944, production of the 76mm (3in)-gunned T-34 was gradually phased out in favour of the up-gunned T-34/85. According to some sources the final batches of T-34/76 tanks (as they were then often called to differentiate them from the T-34/85) were designated the T-34 Model 1944 (or T-34G according to the Germans). These included a wireless radio set mounted inside the turret and improvements to the commander's gun sights.

T-34/85 Medium Tank

From December 1943, the Soviets incorporated the deadly long-barrelled 85mm gun into their battle-proven T-34 design, ensuring that they had a medium tank of sufficient lethality, survivability and mobility to complete the final defeat of the Axis invaders.

The combat experiences gained by the Soviets during the July 1943 Battle of Kursk had shown them that the 76mm (3in) F-34 gun mounted in the T-34 was no longer adequate against the latest generation of potent German tanks – the devastating Panther and the Tiger. Even though the Soviets were loath to disrupt existing T-34 production rates, the High Command nonetheless recognized that they had to up-gun the existing 76mm (3in)-equipped T-34 Model 1943 design if this vehicle was to retain its combat efficiency in the forthcoming battles intended to drive the hated Axis invaders westward out of Soviet territory.

During autumn 1943 Soviet teams developed four rival 85mm (3.35in) gun designs and two redesigned/enlarged single-piece cast three-man T-34 turrets; incorporation of a three-man turret crew thus increased the tank's crew from four to five. Both turrets had increased armour that raged from 52mm (2.05in) on the sides to 90mm (3.54in) on the front. After trials the Soviets deemed the ZiS-53 gun to be the most suitable for mass-production in the new T-34 design. Initially it proved impossible, however, to mount this weapon in the newly-developed three-man T-34 turret that had been produced at the Krasnoye Sormovo Factory #112 at Gorky; several technical concerns with the ZiS-53 gun had also to be addressed.

Rather than have the introduction of an up-gunned T-34 tank delayed

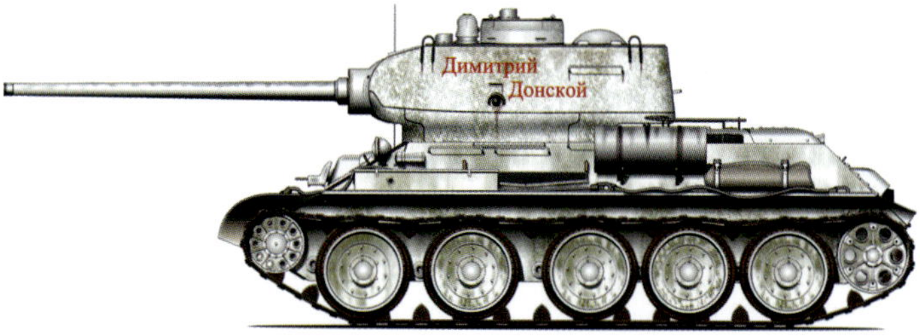

T-34/85 Model 1943

A T-34/85 Model 1943, with inverted U-shaped turret lift rings, roughly finished in two-tone winter camouflage.

by these technical difficulties, the Soviets decided to rush straight into mass production even though no working prototype of the new tank yet existed. Instead, as an interim measure in late December 1943 Factory 112 commenced T-34/85 Model 1943 production. This tank married the existing T-34 chassis (with frontal armour increased to 60mm/2.4in) to the new Factory 112-designed three-man turret that mounted the 85mm (3.35in) D-5T gun.

The first tanks completed reached elite Guards armoured formations during March 1944. This 32 tonne (35.3 ton) vehicle featured a new enlarged curved-sided 'egg-shaped' round-edged turret mounted well-forward on the hull superstructure roof. The combination of the forward turret location and the longer-barrelled gun meant that the main armament projected 2.05m (6ft 9in) beyond the front of the tank.

D-5T gun

The 51.6 calibres-long D-5T was a modified anti-tank version of the combat-proven long-barrelled 52K Model 1939 anti-aircraft gun. The T-34/85 Model 1943's D-5T gun delivered an Armour-Piercing (AP) round at a muzzle velocity of 785m/s (2575 ft/s), enabling it to penetrate 100mm- (3.94in-) thick armour at 1000m (3280ft).

Model 1943 tanks with the D-5T gun sported a distinctive short and wide cylindrical gun mantlet-collar, whereas tanks fitted with the ZiS-S-53 gun had a longer and narrower cylindrical mantlet-collar. Model 1943 turrets also had pistol ports and vision slits on both sides as well as distinctive inverted-U lift rings.

The addition of a third turret crew member and the redesigned positions improved the tactical ergonomics of the T-34/85. The commander was now located further back at the left rear of

T-34/85 Model 1944

Crew 5
Production 1944–45
Weight 32 tonnes (35.3 tons)
Dimensions 6.1m (20ft) x 3m (9ft 10in) x 2.72m (8ft 11in)
Engine 373kW (500hp) Model V-2-34 diesel
Road Speed 55km/h (34mph)
Range 350km (217 miles)
Armament 1 x 85mm (3.35in) ZiS-53 or ZiS-S-53 L/51.5 gun; 2 x 7.62mm (0.3in) DT machine-gun, typically co-axial
Armour 20–90mm (0.79–3.54in)

T-34/85 Model 1944

This T34/85 features the 'flattened' turret, ZIS-S-53 gun and improvised screens designed to protect against panzerfaust charges during the assault on Berlin, April 1945. Note the full disc road-wheels, which was unusual for a late production model.

the turret (with his turret-roof armoured cupola accordingly moved); the gunner was positioned forward of the commander on the left with the loader in the front right.

The reduction of the burden on the tank commander increased the command and control capabilities he could exert in tactical situations. In total, Soviet factories produced around 300 T-34/85 Model 1943 tanks.

Model 1944

During early 1944, as D-5T-equipped T-34/85 Model 1943 tanks came off the production lines, Soviet design teams modified the ZiS-53 gun to create the 51.5 calibres-long ZiS-S-53. Starting in spring 1944, increasing numbers of T-34/85 Model 1944 tanks came off the production lines (particularly at Factory 183) mounting either the 85mm (3.35in) ZiS-53 or the modified ZiS-S-53 gun instead of the D-5T.

Firing Armour-Piercing High-Explosive (APHE) rounds at a muzzle velocity of 792m/s (2598ft/s), the T-34/85's ZiS-S-53 gun could penetrate 102mm (4.02in) thick armour at 1000m (3280ft). The T-34/85 carried 56 main gun rounds, significantly fewer than the T-34/76. During 1944, Soviet factories

Above: T-34/85 tanks of the Third Guards Tank Army wait in a woodlands close to Berlin, April 1945.

This tank is from the 25th Tank Brigade. The turret inscription indicates it is from 2nd Company, 3rd Platoon, 5th vehicle.

manufactured 11,800 T-34/85 Model 1944 tanks with the S-53 or ZiS-S-53 guns, at a monthly average of around 983 tanks.

Model 1944 tanks featured two main turret designs – the 'standard' turret (vehicles produced by Factories 112, 174 and 183) and the 'flattened' turret (built by Factory 183). The latter was so named because the slight bulge in the turret side's bottom rear had been flattened. Model 1944 turrets also had a long narrow cylindrical gun mantlet-collar, the commander's cupola moved further to the rear, a view port only on the right-hand side and the radio antennae for its radio fitted to its roof (rather than on the hull top as in the Model 1943).

Model 1944 tanks that featured the S-53 gun had a long, thin mantlet-collar that bulged at the sides, whereas the similar mantlet-collar for the ZiS-S-53 had no bulges. Further minor design modifications, including an enlarged commander's cupola, were incorporated into the Model 1945 variant. By the war's end in May 1945, 23,100 T-34/85s of all types had been produced at a monthly average of over 1400 vehicles.

T-34/85 Model 1944

This cutaway of the T-34/85 illustrates some of the key internal components of this complex weapons system.

TARGETING TECHNICALITIES

In the T-34/85 Model 1944 tank fitted with the 85mm (3.35in) ZiS-S-53 gun, this 51.5-calibre long main armament could be manually elevated through a range from -5 degrees through to 22 degrees. The ZiS-S-53 gun also sported a TSh-16 Mark IV telescopic sight. To assist in acquiring targets, the turret was electrically traversed, taking 21 seconds to complete a 360-degree turn. The vehicle carried 56 85mm (3.35in) rounds.

T-34/85 Model 1944

This mid-1944 T-34/85, built in Zavod No.112 'Krasnoye Sormovo', features the 85mm ZIS-S-53 gun in an unusual mantlet.

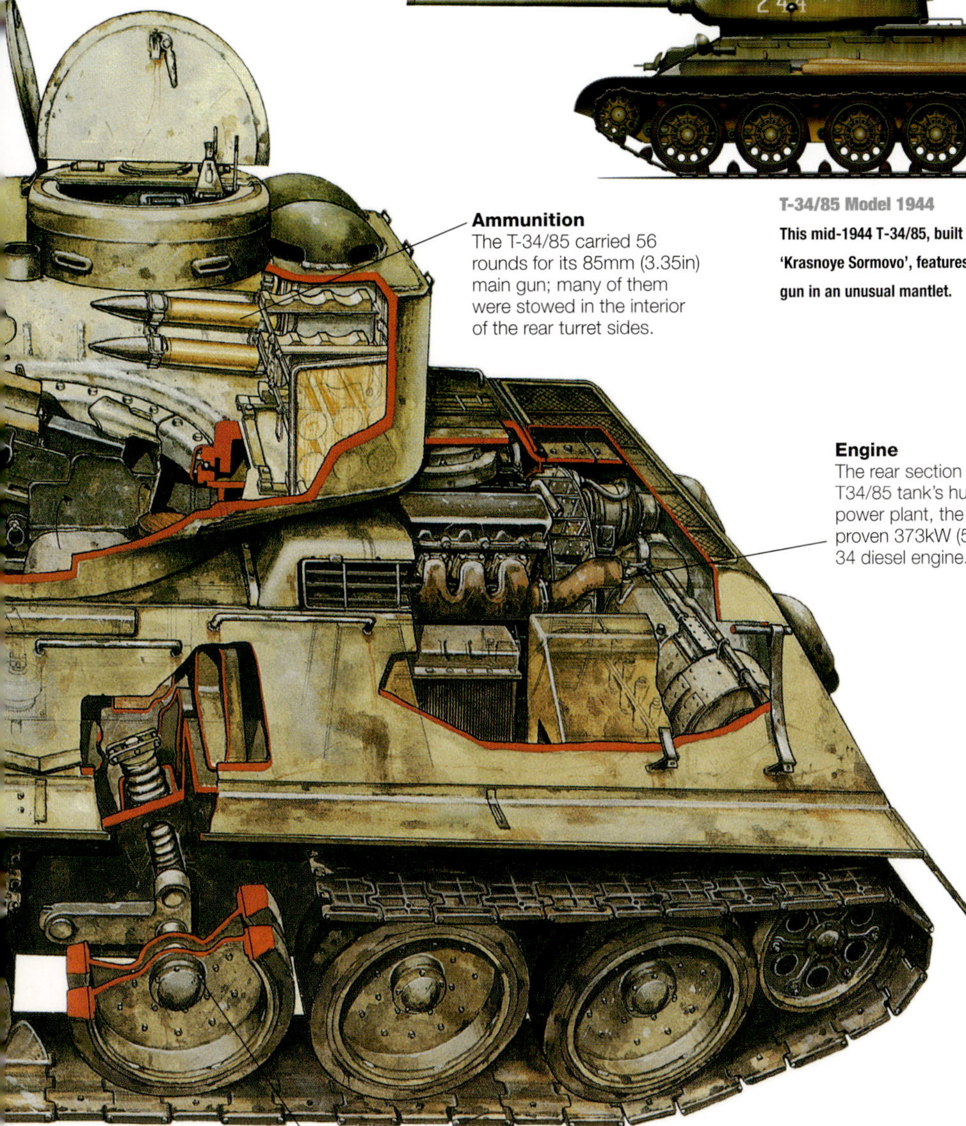

OT-34 Flame Tank

As per previous practice – such as with the T-26 light tank – the Soviets utilized the combat-proven T-34 design as the basis for several specialized vehicles, including the OT-34 flame-throwing tank.

During 1942, the Soviet converted T-34 Model 1942 tanks into flame-throwing variants that deployed a new flame weapon, the ATO-41, which had a range of 90m (300ft). This stubby-barrelled device was fitted to the T-34's right hull front in place of the standard machine-gun, which forced the wireless radio to be moved to the turret.

This four-man variant retained the turret-mounted 76.2mm (3in) F-34 gun and co-axial machine-gun mounted in the T-34 tank. The OT-34 had 105 litre (23 gal) fuel tanks that were housed internally, which made it cramped for the crew to operate within. These tanks enabled the flame device to deliver ten four-to-five second jets of flame.

In comparison with the standard T-34 Model 1942 battle tank, the modifications incorporated into the OT-34 raised the vehicle's weight by several tonnes to 31 tonnes (34.2 tons).

OT-34
Crew 4
Production 1942–43
Weight 31 tonnes (34.2 tons)
Dimensions 5.92m (19ft 5in) x 3m (9ft 10in) x 2.45m (8ft)
Engine 373kW (500hp) Model V-2-34 diesel
Road Speed 55km/h (34mph)
Range 465km (289 miles)
Armament 1 x 76.2mm (3in) F-34 L/40.5 gun; 1 x ATO-41 flame device; 1 x 7.62mm (0.3in) co-axial DT machine-gun
Armour 20–70mm (0.79–2.76in)

Improved device

During 1943–44 an improved device, the ATO-42, was fitted both to T34 Model 1943 tanks, which were also designated OT-34, or to T-34/85 tanks, and designated OT-34/85. The ATO-42 had an enhanced maximum effective range of 130m (430ft) while the enlarged internal fuel tanks allowed for an additional five bursts of flame.

Factory #112 produced 1170 OT-34 and 210 OT-34/85 flame tanks before the war's end in May 1945.

FLAME TANK DEVELOPMENT

Soviet flame-throwing tank development occurred in two main phases. First, during the 1930s the Soviets led the world in the development of such technology, using the T-26 light tank design as the basis of the KhT-26, KhT-130 and KhT-133 flame-throwing tanks. The interest in such vehicles was re-kindled during 1942. In this second phase the Soviets developed the OT-34 flame tank as well as the KV-8, which was based on the KV-1 heavy tank design.

OT-34 flame-throwing tank

The conical short-barrelled hull-front flame device identifies this as an OT-34, converted from a T-34 Model 1943, as evidenced by the drum-like commander's cupola.

T-44 Medium Tank

Entering service in early 1945, the advanced T-44 medium tank further improved and simplified the T-34/85 design and paved the way for the introduction in 1947 of the famous T-54 tank.

During 1944–45, Soviet designers picked up the T-34M development work (investigating an improved T-34/76) that had been abandoned in 1941. Initially the 1944 developmental work aimed at producing an improved T-34/85 design, but this morphed into the development of a new variant, designated the T-44 medium tank.

The 31.9 tonne (35.2 ton) vehicle mounted the 85mm (3.35in) D-5T gun (as fitted to the T-34/85 Model 1943) on a redesigned and up-armoured T-34/85 turret. The T-44 also had a redesigned and shallower chassis, optimized for mass production, with its engine re-orientated onto a novel transverse mounting.

The T-44 abandoned the Christie coil-spring suspension in favour of a more economic torsion bar arrangement that supported five large road-wheels per side. The ensuing design had a lower silhouette than its predecessor, as well as sleek well-sloped lines.

T-44
Crew 4
Production 1945
Weight 31.9 tonnes (35.2 tons)
Dimensions 7.65m (25ft 1in) x 3.15m (10ft 4in) x 2.45m (8ft)
Engine 373kW (500hp) Model V-44 diesel
Road Speed 51km/h (32mph)
Range 300km (186 miles)
Armament 85mm (3.35in) D-5T gun; 2 x 7.62mm (0.3in) DTM machine-gun, 1 co-axial, 1 hull front
Armour 15–120mm (0.59–4.72in)

Enhanced model

In comparison to the T-34/85, the T-44 featured enhanced protection that ranged from 15mm (0.59in) to 120mm (4.72in) thickness. The turret front was protected by 120mm (4.72in) armour, 25 per cent more than its predecessor, and the hull glacis by 90mm (3.54in) of armour – almost double that of the T-34/85.

Powered by a 373kW (500hp) Model V-44 diesel engine, the T-44 could achieve a top road speed of 51km/h (32mph) and a maximum operational range of 300km (186 miles).

T-44 medium tank
The long and shallow turret of the T-44, together with its shallow hull superstructure, heavily influenced post-war Soviet tank design.

About 175 T-44 medium tanks were delivered to front-line units before the war's end in May 1945. Probably the world's most advanced medium tank at that time, the design formed the basis of the famous post-war Soviet medium tank, the 100mm- (3.94in-) gunned T-54.

THE T-44-100 PROTOTYPE
During the T-44's development the designers investigated mounting the more powerful 100mm (3.9in) gun in the new design. When the gun was mounted on the T-44-100 two problems became apparent. First, the long gun fitted to a forward-positioned turret gave the tank a significant barrel overhang. Second, the gun's large breech made the interior of the turret cramped. These experiments, however, did lead to the creation of the T-54 in 1947.

HEAVY TANKS

Throughout the bitter struggle on the Eastern Front during 1941–45, the mighty heavy tanks the Soviets produced, such as the KV-1 and IS-2, provided the lethality and survivability required to drive back the Axis invaders.

This chapter features the following tanks:

- T-35 Heavy Tank
- KV-1 Heavy Tank
- KV-1S Heavy Tank
- KV-85 Heavy Tank
- IS-1 Heavy Tank
- IS-2 Heavy Tank
- IS-3 Heavy Tank

An IS-2 tank moves into a German town. The heavy, turret-mounted DShK anti-aircraft machine gun has been removed and replaced by the DT 7.62mm (0.3in) weapon with a flat pan magazine. The DT machine gun was much less cumbersome for use in urban warfare.

T-35 Heavy Tank

The first Soviet attempt to produce an effective indigenous heavy tank design – the five-turreted T-35 behemoth – produced a visually impressive vehicle ideal for propaganda purposes, yet one that was tactically disappointing.

During 1930–33, the Soviets designed their first heavy tank, being influenced by the T-28 and the Vickers Independent. During 1933–34 the Soviets manufactured 20 pre-production T-35 tanks, but the Model 1936 main production version only entered service in 1936.

The T-35 was a long 43 tonne (47.4 ton) multi-turreted tank. The vehicle's main turret, mounted high above the central hull superstructure, sported a 76.2mm (3in) Model 1927/32 gun. Surrounding the main turret and located beneath it were four auxiliary turrets, with two positioned forward and two aft of the main turret. These auxiliary turrets mounted either a 45mm (1.77in) 20K Model 1932 gun or a 7.62mm (0.3in) DT machine-gun. To man these five turrets, the T-35 needed an 11-man crew.

Despite its weight and length – 9.72m (31ft 11in) – the T-35 was only modestly protected by 11–30mm (0.43in–1.18in) armoured plates. During 1935–1938 the Soviets manufactured 35 T-35 Model 1936 tanks. During 1938–39 six modified T-35 Model 1939 tanks were produced, which featured sloped turret side armoured plates.

In service the T-35 was disappointing. The tank was hard to manoeuvre and it proved difficult to employ the five weapons accurately when in motion. The T-35s that were completed were mainly used for propaganda: these behemoths made a fine show during the annual May Day parades. The few T-35s that entered combat during 'Barbarossa' were soon either knocked out or abandoned.

T-35 (Model 1936)

This side view illustrates the T-35 tank's great length and the complex arrangement of its five turrets.

T-35 (Model 1939)

This modified variant incorporated sloped turret side armoured plates as well as a roof-mounted anti-aircraft machine-gun.

T-35 (Model 1936)

Crew 11

Production 1935–39

Weight 43 tonnes (47.4 tons)

Dimensions 9.72m (31ft 11in) x 3.2m (10ft 6in) x 3.43m (11ft 3in)

Engine 373kW (500hp) M-17M petrol

Road Speed 30km/h (19mph)

Range 150km (93 miles)

Armament 1 x 76.2mm (3in) Model 1927/32 gun; 2 x 45mm (1.77in) 20K Model 1932 gun; 2 x 7.62mm (0.3in) DT machine-gun

Armour 11–30mm (0.43in–1.18in)

KV-1 Heavy Tank

During Operation 'Barbarossa', the occasional counter-attacks made in strength by the potent new Soviet heavy tank design, the KV-1 – supported by T-34s – inflicted considerable shock on the Axis invaders.

KV-1 Model 1939

The large turret and shallow hull superstructure of the KV-1 are evident in this side view, as is the Model 1939's distinctive gun mantlet.

KV-1 Model 1939
Crew 5
Production 1939–40
Weight 43.6 tonnes (48.1 tons)
Dimensions 6.68m (21ft 11in) x 3.32m (10ft 11in) x 2.71m (8ft 11in)
Engine 473kW (600hp) V-2 12-cylinder diesel
Road Speed 35km/h (22mph)
Range 160km (99 miles)
Armament 1 x 76.2mm (3in) L-11 L/32 gun; 3 x 7.62mm (0.3in) DT machine-gun, 1 co-axial, 1 turret rear, 1 hull front
Armour 25–90mm (0.98–3.54in)

KV-1 Model 1939

During 1938–40 Soviet engineers designed a new five-man heavy tank, the Klimenti-Voroshilov (KV)-1 Model 1939. One KV-1 prototype performed well during the 1939–40 Winter War with Finland, and subsequently pre-production commenced in late 1939. The Model 1939 was a shallow-hulled vehicle with a large, tall, sloped-sided welded turret that mounted the short-barrelled (32 calibres-long) 76.2mm (3in) L-11 gun. The tank also sported three 7.62mm (0.3in) DT machine-guns.

Weighing 43.6 tonnes (48.1 tons), the KV-1 was well protected with 75mm (2.95in) armoured plates on the hull front and 90mm (3.54in) turret front plates. Despite its weight, the KV-1 was reasonably mobile, thanks to a decent ground pressure figure generated by the tank's wide tracks; its maximum road speed was 35km/h (22mph). Its suspension/track arrangement consisted of six medium-sized solid-disc road-wheels per side, each connected to a torsion-bar suspension, with a rear driving wheel and a front idler. Powered by the 473kW (600hp) V-2 12-cylinder diesel engine, the KV-1 could obtain a maximum by-road operational range of 160km (99 miles). The Model 1939, however, was plagued by its unreliable transmission and poor crew ergonomics. During 1940 some 141 Model 1939s were produced.

KV-1 Model 1940

During 1940 modifications were incorporated into the KV-1 design to produce the Model 1940. The most obvious external difference between

KV-1 Model 1940

Note the redesigned mantlet of the F32-equipped late-production Model 1940. This model with three-tone armour fought on the Leningrad Front in spring 1942.

KV-1 Model 1940
Crew 5
Production 1940–41
Weight 43.6 tonnes (48.1 tons)
Dimensions 6.68m (21ft 11in) x 3.32m (10ft 11in) x 2.71m (8ft 11in)
Engine 473kW (600hp) V-2 12-cylinder diesel
Road Speed 35km/h (22mph)
Range 335km (208 miles)
Armament 1 x 76.2mm (3in) L-11 L/32 or F-32 gun; 3 x 7.62mm (0.3in) DT machine-gun, 1 co-axial, 1 turret rear, 1 hull front
Armour 25–75mm (0.98–2.95in)

APPLIQUÉ ARMOUR

The up-armouring of existing designs through the incorporation of bolted-on appliqué armour was a short-term expedient employed by many belligerents during the Second World War. This technique was a cheap and quick way to increase the survivability of a tank design in the face of increasing enemy anti-tank threats. It was, however, still an expedient as such vehicles remained less survivable than new designs that incorporated single-plate armoured protection of the same thickness as the vehicles augmented with appliqué plates.

KV-1E

The slogan on the side of the turret reads 'Victory will be ours.'

this model and its predecessor, the Model 1939, was that they each had a differently-shaped gun mantlet. The Model 1939 had the distinctive low-slung 'boar's head' mantlet, with space for the gun recoil above the barrel. Late-production Model 1940s, however, instead mounted the 76.2mm (3in) F-32 gun; this had the recoil mechanism under the barrel, so the vertical bottom face of the box-like gun mantlet fitted beneath the barrel. Firing BR-350A Armour-Piercing (AP) rounds, the F-32 gun had a maximum muzzle velocity of 612m/s (2007ft/s), and could penetrate 57mm (2.24in) of armour at a range of 1000m (3280ft).

The other external feature that differentiated the Model 1940 from its predecessor was that it lacked the two canisters per side of lubricant that had been fitted on the Model 1939's rear fenders. Internally, the Model 1940 was powered by the same 473kW (600hp) V-2 diesel engine as the Model 1939. The Kirovskiy factory (LKZ) at Leningrad produced 243 Model 1940 tanks.

KV-1E

The Soviet experience of employing the KV-1 to resist Operation 'Barbarossa' during 1941 showed that while it initially seemed impregnable to German anti-tank weapons, this superiority gradually declined as the enemy adjusted their tactics by using the potent 88mm (3.46in) flak gun in an anti-armour role. During late 1941 and 1942, around 150 KV-1 Model 1939, 1940 and 1941 tanks were dispatched back to the KTZ factory for survivability augmentation work to create the KV-1E design; the 'E' stood for 's. Ekranami' or 'with skirted armour.' At KTZ workers bolted large soft steel 20mm (0.79in) or 30mm (1.18in) appliqué armoured plates to the turret front and sides, as well as the hull front. These plates increased the KV-1E's frontal protection to 110–120mm (4.33–4.72in) thickness, which once again rendered the tank invulnerable to German direct-fire weapons.

Inevitably, the additional bolted-on armour increased the vehicle's weight significantly to 48.95 tonnes (53.95 tons). This additional weight further reduced the tank's modest

KV-1E
Crew 5
Production 1941–42
Weight 48.95 tonnes (53.95 tons)
Dimensions 6.68m (21ft 11in) x 3.32m (10ft 11in) x 2.71m (8ft 11in)
Engine 473kW (600hp) V-2 12-cylinder diesel
Road Speed 34km/h (21mph)
Range 335km (208 miles)
Armament 1 x 76.2mm (3in) L-11 L/32 or F-32 gun; 3 x 7.62mm (0.3in) DT machine-gun, 1 co-axial, 1 turret rear, 1 hull front
Armour 25–75mm (0.98in–2.95in) plus 20–30mm (0.79–1.18in) appliqué plates

KV-1E

The large round bolt-heads on the turret sides of this KV-1E identify where the additional appliqué armoured plates have been affixed to the existing armour.

KV-1 Model 1941

Sporting new radial road-wheels, the Model 1941 typically mounted either the longer-barrelled F-34 or ZiS-5 gun in a redesigned and up-armoured welded turret with modified mantlet. This ChTZ-produced tank is named 'Merciless', and was heavily decorated by famous Moscow artists in early 1942.

mobility as well as strained its already inadequate transmission. As a result, the KV-1E's top road speed fell slightly to 34km/h (21mph).

KV-1 Model 1941

During late 1941 a series of incremental design modifications were incorporated into the KV-1 Model 1940 design to create a variant that was retrospectively referred to as the Model 1941. Perhaps the most obvious external difference discernible in the Model 1941 in comparison with its predecessors was that its six road-wheels per side featured a new radial-style solid-disc design. The most significant modification, however, was that most KV-1 Model 1941 tanks incorporated a more potent main armament, the newly-developed long-barrelled 76.2mm (3in) F-34 or ZiS-5 gun. This 40.5 calibre-long weapon was identical to the one mounted in the T-34 Model 1941 medium tank. Having the same main armament in both these designs permitted some simplification of the overall Soviet tank production effort. When firing Armour-Piercing High Explosive (APHE) rounds, the KV-1 Model 1941's F-34 gun could penetrate 60mm (2.36in) of vertical armour at the range of 1000m (3280ft).

KV-1 Model 1941

This vehicle is an early-production Model 1941 with the old F-32 gun and mantlet fitted to the new up-armoured welded turret.

KV-1 Model 1941

Two F-34-equipped Model 1941 tanks, both with cast turrets. The top has a fading whitewash scheme with rare cupola-mounted machine-gun, the lower a monochrome camouflage.

KV-1 Model 1941
Crew 5
Production 1941–42
Weight 45 tonnes (49.6 tons)
Dimensions 6.75m (22ft 2in) x 3.32m (10ft 11in) x 2.71m (8ft 11in)
Engine 473kW (600hp) V-2 12-cylinder diesel
Road Speed 35km/h (22mph)
Range 335km (208 miles)
Armament 1 x 76.2mm (3in) F-34 or ZiS-5 L/40.5 gun; 3 x 7.62mm (0.3in) DT machine-gun, 1 co-axial, 1 turret rear, 1 hull front
Armour 30–90mm (1.18–3.54in)

The longer barrel, which extended a modest 100mm (3.94in) beyond the front of the vehicle hull, was the main external feature that identified most Model 1941 tanks from their predecessors, which mounted the shorter-barrelled 76.2mm (3in) L-11 or F-32 weapons. Many of the first 200 or so KV-1 Model 1941 tanks, however, still mounted the old F-32 gun, due to shortages in the delivery of the new weapon.

The KV-1 Model 1941 tank also incorporated a redesigned and up-armoured welded or cast turret. This can be identified from earlier turrets because its bottom edge ran in a straight line from front to back, rather than cutting upwards then backwards where the box-like rear stowage bin began to project from the back of the turret circle. The Model 1941 also featured a slightly modified box-like gun mantlet, similar to that of the Model 1940, which had been optimized to the shape of the new F-34 or ZiS-5 gun. Other minor design modifications were incorporated to speed up the manufacturing process. During late 1941 and throughout 1942, ChTZM (the amalgamated LTZ Leningrad factory plus the ChTZ plant) at Chelyabinsk's 'Tank City' produced 1200 Model 1941 tanks.

KV-1 Model 1942

Production of the final KV-1 variant, the Model 1942, overlapped with that of the final batches of Model 1941 tanks during the first eight months of 1942. The Model 1942's principal innovation was the incorporation of a thickened reinforced cast turret. This was externally almost indistinguishable from the cast turret of the Model 1941. However, the bottom fringe of the turret flanged outward more than in the previous turret, while the rear turret machine-gun ball-mount featured a new armoured collar. The new turret had its armour thickness increased to 120mm (4.72in) in certain key areas of the cast design, although the turret's overall weight was not increased. The hull superstructure also had its armour thickened in key places, which led to this variant being even heavier than its predecessor.

With the Model 1942's weight increased to 48 tonnes (52.9 tons), the already over-stressed engine, transmission and suspension was placed under even greater strain. This

KV-1 PRODUCTION

During a four-year production run lasting from 1939 through to 1942, Soviet factories produced 3350 KV-1 heavy tanks and converted 150 of these to KV-1E 'screened-armour' variations. The manufacturing run delivered 141 Model 1939 vehicles, 243 Model 1940s, 1200 Model 1941 vehicles and 1700 Model 1942s. Initially the KTZ factory at Leningrad monopolized KV-1 production, but from 1942 several other relocated factories joined in, particularly the merged ChTZM works at 'Tank City' in Chelyabinsk to the east of the Ural Mountains.

led Model 1942 tanks frequently to require repairs or replacement parts for these components. The increased weight also reduced the Model 1942's top road speed to a mere 28km/h (17mph), while its by-road operational range declined to 250km (155 miles) from its predecessor's 335km (208 miles). By late 1942, while the KV-1 remained heavily protected and well-armed, the design's mobility, reliability and ergonomics had become mediocre in comparison to the newest German tank designs.

Although the Model 1942 mounted the same ZiS-5 gun as its predecessor, main gun round stowage was increased from 111 to 114 rounds. Most Model 1942 tanks also had an anti-aircraft mount added to the top of the commander's cupola on the turret roof. This housing mounted the standard 7.62mm (0.3in) DT machine-gun, which increased the number of machine-guns in the KV-1 design from three to four.

The early KV-1 Model 1942 tanks essentially retained the same hull as its predecessor, with rounded hull rear plates. The late-production Model 1942 run, however, incorporated a slightly redesigned hull. This was identified by its flat rear plate, which sloped outwards at a 50-degree angle. In total, Soviet factories produced some 1700 KV-1 Model 1942 tanks during 1942.

KV-1 Model 1942
Crew 5
Production 1942
Weight 48 tonnes (52.9 tons)
Dimensions 6.8m (22ft 4in) x 3.32m (10ft 11in) x 2.71m (8ft 11in)
Engine 473kW (600hp) V-2 12-cylinder diesel
Road Speed 28km/h (17mph)
Range 250km (155 miles)
Armament 1 x 76.2mm (3in) F-34 or ZiS-5 L/40.5 gun; 4 x 7.62mm (0.3in) DT machine-gun, 1 co-axial, 1 turret rear, 1 turret roof, 1 hull front
Armour 30–120mm (1.18–4.72in)

Model 1942
The Model 1942 design typically featured a reinforced cast turret and a cupola-mounted machine-gun.

KV-1S Heavy Tank

In an attempt to improve the lumbering manoeuvrability of the up-armoured KV-1 Model 1942 tank, the Soviets designed a new variant, termed the KV-1S, with lighter armour and enhanced speed.

The up-armouring of the KV-1 – in the form of the KV-1E, Model 1941 and Model 1942 designs – obviously increased the weight of the vehicle. The original KV-1 Model 1939 weighed 43.6 tonnes (48.1 tons), whereas by 1942 it weighed 48 tonnes (52.9 tons), a 10 per cent increase. This extra weight placed further strain on the old-fashioned transmission and the under-powered V-2 diesel engine. As a consequence the KV-1's operational range was diminished from its 1940 peak of 335km (208 miles) down to just 250km (155 miles). Similarly, its already meagre maximum road speed fell from 35km/h (22mph) to 28km/h (17mph); the tank's actual cross-country speed in action was barely half this.

By 1942 the Soviets were finding it increasingly hard to tactically integrate the KV-1 Model 1942 with the existing T-34 Medium Tank squadrons during combat. From October 1942 onwards the Soviets applied one partial sticking plaster to this problem – they concentrated many Model 1942s into separate armoured brigades for use in assault and infantry support missions. The other temporary solutions the Soviets utilized was the design of a lighter, more mobile KV-1 tank; this was designated the KV-1S Model 1942, with the 'S' standing for 'skorostnoy', meaning 'speedy.' By creating a 43.4 tonne (47.8 ton) heavy tank that could move by road at a top speed of 43km/h (26mph), the Soviets hoped that the KV-1S could effectively operate with the T-34 in combat conditions.

The Soviets achieved a more mobile KV-1 variant through the combination of five principal elements: reducing the levels of armour protection in the new variant; reducing the size and weight of the cast turret; reducing the hull weight with a sloping engine deck; reducing the weight of the road-wheels with small holes between the radials; and incorporating an enhanced planetary transmission, gearbox, main clutch, improved engine cooling features and an up-rated 485kW (650hp) V-2K diesel engine.

Modified turret

The KV-1S had a slightly smaller and lighter turret with rounded edges and a rear face that sloped outwards; it also incorporated for the first time a commander's cupola with all-round vision blocks. These modifications aside, the KV-1S mounted the same 76.2mm (3in) ZiS-5 gun as seen in previous variants with main armament stowage remaining at 114 rounds. The last batches of the design – the Model 1943 – featured redesigned road-wheels that returned to the no-holed solid-disc design seen in early KV-1 tanks. These late-production tanks also featured horizontal handrails fitted to the turret rear sides as well as to the hull deck front and rear to assist infantry being carried into combat.

KV-1S

This KV-1S heavy tank is most unusual in having a palm-headed radio antenna on its front hull side decking.

KV-1S

Crew 5
Production 1942–43
Weight 43.4 tonne (47.8 ton)
Dimensions 7.41m (24ft 4in) x 2.87m (9ft 5in) x 3.25m (10ft 8in)
Engine 485kW (650hp) V-2K 12-cylinder diesel
Road Speed 43km/h (26mph)
Range 250km (155 miles)
Armament 1 x 76.2mm (3in) ZiS-5 gun; 4 x 7.62mm (0.3in) DT machine-gun, 1 co-axial, 1 hull front, 1 turret rear, 1 turret roof
Armour 20–82mm (0.79in–3.2in)

KV-1S (winter camouflage)

This side view of a KV-1S from spring 1943 sports a fading winter white camouflage scheme.

KV-1S

These views clearly illustrate the new lighter turret and lightened road-wheels of the KV-1S.

Between August 1942 and August 1943 Soviet factories produced 1370 KV-1S tanks at a monthly average of 114 vehicles; the largest number of KV-1S tanks produced in any one month was 180 in September 1942.

Although a reasonably effective stop-gap the KV-1S was in effect an inferior and costlier-to-produce competitor to the T-34 Medium Tank. The KV-1 design had reached the end of its shelf-life, and with the mid-1943 appearance of the German Panther tank what the Soviets desperately needed was a new, modern, up-gunned, up-armoured and up-powered heavy tank design.

TANK-INFANTRY COOPERATION

In the middle and later phases of the war against the Axis invaders (namely 1942–45), Soviet infantry adopted the tactic of moving into contact with the enemy by standing or sitting on a tank's hull decking. To facilitate these tactics, and reduce the risk of soldiers falling off as the vehicle manoeuvred, tanks such as the KV-1S incorporated horizontal handrails on the turret and hull sides onto which soldiers could hold.

In this propaganda photo, a column of KV-1S tanks of the 6th Guards Armoured Breakthrough Regiment advance to the front line in spring 1943.

KV-85 Heavy Tank

As an interim response to encountering the latest German tanks in the battles of summer 1943, the Soviets mounted a more potent 85mm (3.35in) gun in an enlarged turret on top of the KV-1S to produce the KV-85.

The Soviet experience of encountering the new German Panther Medium Tanks and Tiger heavy tanks during the July 1943 Battle of Kursk led them to conclude that Soviet heavy tanks urgently needed to be up-gunned to counter this new threat. While development work unfolded on the new Iosef Stalin (IS)-1 heavy tank, the four-man KV-85 was produced as an interim stop-gap. This design involved mounting the new 51.6 calibre-long 85mm (3.35in) D-5T gun in an enlarged turret on top of a modified KV-1S hull; this new gun was then being incorporated into the T-34/85 medium tank design. Firing the standard anti-armour round the D-5T gun mounted in the KV-85 could penetrate 100mm (3.94in) of vertical armour at the range of 1000m (3280ft).

New turret

The new turret incorporated into the KV-85 had been originally designed for the IS-85 (IS-1) heavy tank, then still in its development phase. The hull of the KV-85 was essentially a slightly modified KV-1S chassis with its frontal armoured plates increased from 82mm (3.2in) to 110mm (4.33in). However, the additional weight of the larger gun, heavier rounds, enlarged turret and thickened hull frontal armour all combined to push the KV-85 back up to 46 tonnes (50.7 tons) – more than the KV-1 Model 1941. This undermined the very logic of increased mobility that had been behind the introduction of the KV-1S in the first place. Thus the KV-85's top road speed was consequently reduced to a disappointing 35km/h (22mph).

During autumn 1943 Soviet factories produced just 148 KV-85 heavy tanks.

KV-85

This stop-gap tank mounted the new 85mm (3.35in) gun in the enlarged IS-1 turret atop a modified KV-1S chassis.

KV-85
Crew 4
Production 1943
Weight 46 tonnes (50.7 tons)
Dimensions 8.6m (28ft 3in) x 3.25m (10ft 8in) x 2.9m (9ft 6in)
Engine 473kW (600hp) V-2 12-cylinder diesel
Road Speed 35km/h (22mph)
Range 250km (160 miles)
Armament 1 x 85mm (3.35in) D-5T L/51.6 gun; 3 x 7.62mm (0.3in) DT machine-gun, 1 co-axial, 1 turret rear, 1 hull front
Armour 30–110mm (1.18–4.33in)

Elite unit

The KV-85 was used to equip a number of Guards regiments in November 1943, including the 7th, 11th, 15th and 29th Guards Heavy Tank Regiments.

IS-1 Heavy Tank

In place of the interim KV-85, the purpose-designed IS-1 mounted the new 85mm (3.35in) gun in a newly-designed chassis that retained much of the mobility restored by the KV-1S design.

With its final development undertaken at the same time as the KV-85, the four-man IS-85 (subsequently re-designated IS-1) was originally based on the KV-13 prototypes, the first of which was completed back in May 1942.

The first IS-1 prototype was demonstrated in front of Stalin in August 1943. The IS-1 mounted the 85mm (3.35in) D-5T gun in a new enlarged cast turret – the precise combination incorporated into the stop-gap KV-85. The larger turret ring was incorporated into the design by widening the hull superstructure so that it spread over the top of the tracks.

IS-1 VERSUS TIGER

During summer 1943 at the Kubinka testing grounds, the Soviets conducted firing tests using the new 85mm (3.35in) D-5T gun against a captured German Tiger tank. These tests demonstrated that the D-5T could, under test conditions, penetrate the enemy tank's 100mm (3.94in) frontal armour at a range of 1000m (3280ft). In reality, however, the relatively poor quality of Soviet sights, as well as the nature of armoured tactics, meant that it remained extremely rare for IS-1 tanks during winter 1943–44 to knock out a Tiger tank at anywhere near these ranges.

Mobility and speed

The IS-1 chassis combined the best elements of the KV-1S and the KV-13 prototypes. Its improved suspension, transmission and 473kW (600hp) V-2-IS engine ensured that, despite its weight of 43.4 tonnes (47.9 tons), the IS-1 had mobility only moderately inferior to that of the KV-1S. The IS-1 could obtain a maximum road speed of 37km/h (23mph) and a maximum operating range of 150km (93 miles) when travelling by road.

Soon after the first IS-1 tanks were completed in October 1943, however, the Soviet high command decided that the 122mm (4.8in) A-19 gun should become the main armament to be mounted in its heavy tanks. This required the development of a new design (the IS-2) and therefore in January 1944 the production of the IS-1 was terminated after just 200 tanks had been completed.

The new tanks immediately joined the elite Guards Heavy Tank Brigades then spearheading the winter 1943–44 offensives that pushed westward through the Ukraine.

IS-1

The altered wheel design differentiated the IS-1 from the externally not dissimilar KV-85. This tank was part of the 3rd Belorussian Front, which fought in the Baltic in the summer of 1944.

Crew 4
Production 1943–44
Weight 43.4 tonnes (47.9 tons)
Dimensions 8.56m (28ft 1in) x 3m (9ft 10in) x 2.74m (9ft)
Engine 473kW (600hp) V-2-IS 12-cylinder diesel
Road Speed 37km/h (23mph)
Range 150km (93 miles)
Armament 1 x 85mm (3.35in) D-5T L/51.6 gun; 3 x 7.62mm (0.3in) DT machine-gun, 1 co-axial, 1 turret rear, 1 hull front
Armour 30–120mm (1.18in–4.72in)

IS-2 Heavy Tank

The next heavy tank in the Iosef Stalin heavy tank series, the IS-2, was the first such vehicle to mount the truly formidable new 122mm (4.8in) A-19 gun.

IS-2 (early production series)

The first IS-2 tanks, such as this early production series, were equipped with the 122mm (4.8in) A-19 gun, while later ones had the almost identically looking D-25T.

During October 1943 the Soviets concluded that it was no use having heavy tanks that only fielded the same 85mm (3.35in) weapon as the T-34/85 Medium Tank. It was decided that a larger-calibre, more lethal weapon was required. Although the 100mm (3.94in) BS-3 gun had been shown to be potent, it proved costly to produce.

The Soviets instead decided that the corps-level 122mm (4.8in) Field Cannon A-19 Model 1931 should become the main armament to be mounted in its future heavy tanks. This decision, therefore, required the development of a new heavy tank design – and consequently the IS-122 (later renamed IS-2) tank was born.

The four-man IS-2 design incorporated the long-barrelled 122mm (4.8in) A-19 gun into a forward-positioned turret. This gave the tank a very significant barrel overhang. The A-19 was initially fitted with a single baffle muzzle brake, later modified to a double-chambered form. In test conditions, the A-19 could penetrate 160mm (6.29in) of vertical armour at 1000m (3280ft), although in actual combat the mediocre Soviet optics made a successful hit at such range relatively unlikely. Starting in early 1944 new IS-2 tanks instead mounted

IS-2
Crew 4
Production 1943–45
Weight 46 tonnes (50.7 tons)
Dimensions 9.9m (32ft 6in) x 3.09m (10ft 2in) x 2.73m (8ft 11in)
Engine 473kW (600hp) V-2-IS 12-cylinder diesel
Road Speed 37km/h (23mph)
Range 240km (149 miles)
Armament 1 x 122mm (4.8in) A-19 or D-25T L/48 gun; 2 x 7.62mm (0.3in) DT machine-gun, 1 co-axial, one turret rear; sometimes 1 x 12.7mm (0.5in) DShK machine-gun on turret roof
Armour 30–120mm (1.18–4.72in)

IS-2

This IS-2 includes the D-25T main gun and a 12.7mm (0.5in) DShK heavy anti-aircraft machine-gun on the commander's cupola.

IS-2 'Polar Bear'

This mid-production vehicle, completed in December 1944, is seen here in its late April 1945 camouflage, with 'Polar Bear' insignia, during the battle of Berlin.

the 122mm (4.8in) D-25T gun, which had marginally greater penetrative capability as well as a slightly faster rate-of-fire.

Modified turret

The well-sloped and rounded modified turret narrowed toward the gun mantlet, presenting a smaller area for the enemy to strike. In addition to its main armament, the IS-2 initially mounted two 7.62mm (0.3in) DT machine-guns, with one co-axial and one in the turret rear. Subsequently, a 12.7mm (0.5in) DShK heavy machine gun was fitted upon an anti-aircraft mount on top of the commander's cupola. To limit the scale of weight increase, the IS-2 only carried 28 rounds for its main gun, which proved a significant tactical liability. Both the A-19 and D-25T guns also had a mediocre rate of fire – at most two rounds-per-minute – which also hampered the tactical efficiency of the IS-2.

Hull armour

The IS-2 hull featured improved well-sloped and rounded – but slightly less thick – armour that still provided greater survivability than the thicker-plated KV-85 and IS-1. This was essential to keep the scale of weight increase to tolerable levels, as the gun and its rounds were significantly heavier than the equivalent elements in the KV-1, KV-85 and IS-1 heavy tanks; the IS-2 weighed 46 tonnes (50.7 tons). However, experience with the IS-2 showed that its armoured plates had a disconcerting tendency to splinter when hit, causing casualties inside the tank.

The IS-2 had a standard torsion-bar suspension that supported six independent pairs of road-wheels per side; the minor visual difference was that the top track was slightly undulating as it got pulled over the three high-positioned return rollers. Powered by a 473kW (600hp) V-2-IS diesel engine, the IS-2 could achieve a top road speed of 37km/h (23mph) and a maximum by-road range of 240km (149 miles).

Soviet factories produced 3854 IS-2 tanks during a production run that lasted from December 1943 until summer 1945. The design underwent various modifications during its production run. Mid-production vehicles (the Model 1944) had frontal armour tempered to a greater hardness, although this was a costly manufacturing process. From May 1944, such vehicles also featured a modified straightened welded or cast hull. Modifications were also made to the configuration of the hull bow plate.

THE 'POLAR BEARS'

The 7th Independent Guards Heavy Tank Brigade, part of the Eighth Guards Army, participated in the battle for Berlin with this unusual 'Polar Bear' camouflage scheme. The brigade had got this informal unit nickname from its participation in the snow-clad 1944 Karelian campaign against Finnish forces. Their insignia was a white polar bear surmounted upon a Red Star.

The featured tank also includes the broad white recognition stripe around the turret and across its roof, which was added to Red Army AFVs to reduce 'friendly fire' incidents during the assault on the German capital.

IS-3 Heavy Tank

Soviet wartime heavy tank development culminated in the IS-3, which was not only the most visually striking heavy tank of the entire war, but probably the most powerful as well.

The summer 1944 appearance of the German King Tiger heavy tank on the Eastern Front led the Soviets to design a new variant, the IS-3, which married a new turret to a redesigned IS-2 chassis; this resulted in the most visually-striking tank of the war. To increase the vehicle's survivability without increasing its weight, the four-man IS-3 incorporated novel armoured features.

The innovative cast turret was a shallow and wide well-rounded hemisphere, designed to maximize shot deflection. The IS-3 mounted the well-proven 122mm (4.8in) D-25T gun in a thick gun mantlet that fitted cleanly into the turret's rounded sides. The novel hull front featured two sloping 110mm (4.33in) armoured plates that met in the middle to create the design's infamous 'pike nose', which again maximized shot deflection.

The hull side armour was also well-sloped inwards, meaning the turret rim projected beyond the hull sides, a reality partially obscured by locating two external cylindrical stowage bins on each upper rear hull side. The 45.77 tonne (50.45 ton) IS-3 was powered by the 473kW (600hp) V-2-IS diesel engine and could obtain a maximum road speed of 40km/h (25mph). Production commenced in March 1945 and the first 29 completed vehicles reached the front line during the war's very last days.

Engine
The IS-3 was powered by the same rear-located 473kW (600hp) V-2-IS diesel engine as its two predecessors, the IS-1 and IS-2 heavy tanks.

IS-3

This three dimensional cut-away of the four-man IS-3 heavy tank gives a good idea of the vehicle's sleek exterior lines, as well as key parts of its interior. Visible inside are the driver's position in the hull front, the breach mechanism for the potent 122mm (4.8in) gun and the rear-located engine housing.

Anti-aircraft gun
In addition to co-axial and hull-front 7.62mm (0.3in) DT machine-guns, the IS-3 also mounted a 12.7mm (0.5in) DShK heavy machine-gun in an anti-aircraft mounting on the turret roof.

Main gun
Seen here is the 122mm (4.8in) D-25T gun's breach mechanism; the low-silhouetted turret in which it was mounted restricted both the gun's angle of depression as well as working crew space.

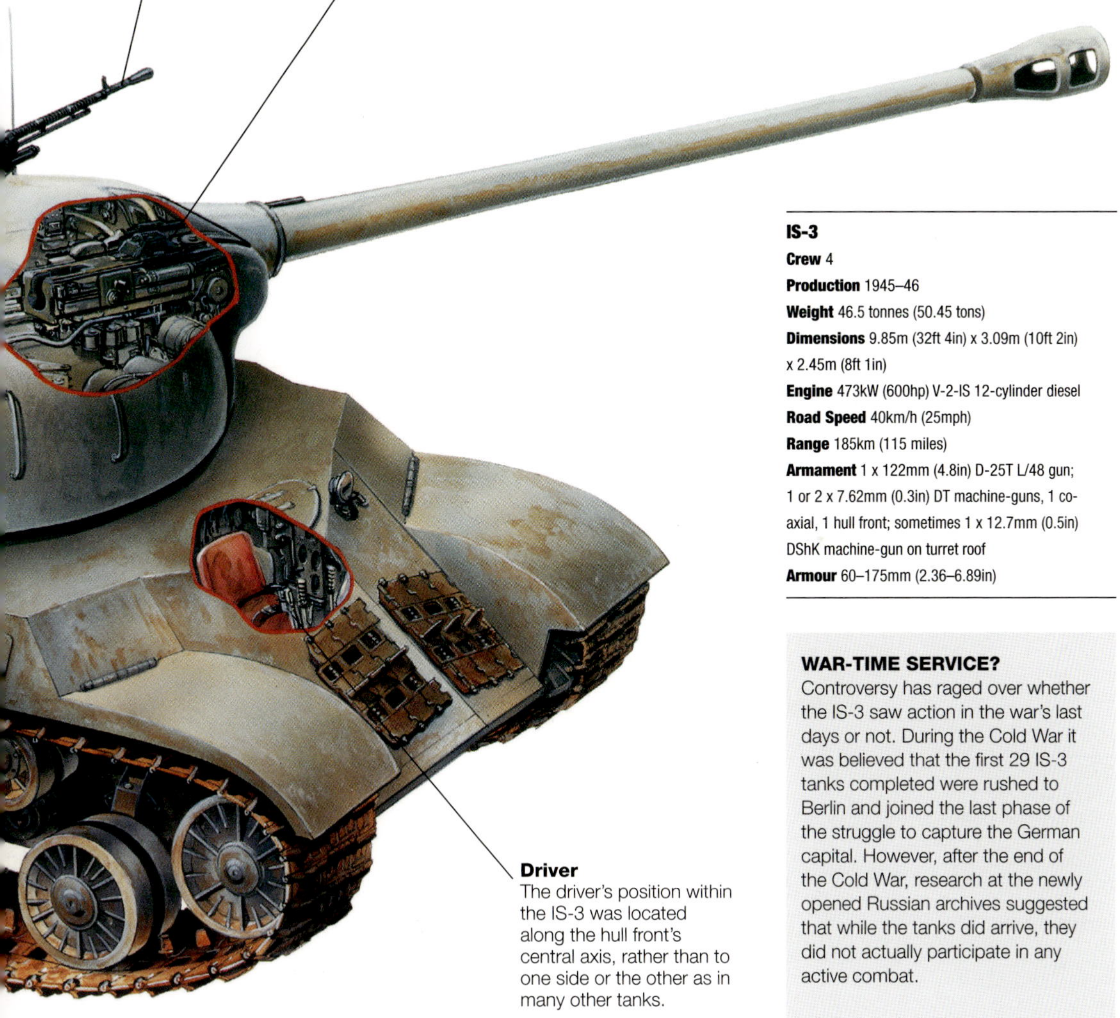

Driver
The driver's position within the IS-3 was located along the hull front's central axis, rather than to one side or the other as in many other tanks.

IS-3

Crew 4

Production 1945–46

Weight 46.5 tonnes (50.45 tons)

Dimensions 9.85m (32ft 4in) x 3.09m (10ft 2in) x 2.45m (8ft 1in)

Engine 473kW (600hp) V-2-IS 12-cylinder diesel

Road Speed 40km/h (25mph)

Range 185km (115 miles)

Armament 1 x 122mm (4.8in) D-25T L/48 gun; 1 or 2 x 7.62mm (0.3in) DT machine-guns, 1 co-axial, 1 hull front; sometimes 1 x 12.7mm (0.5in) DShK machine-gun on turret roof

Armour 60–175mm (2.36–6.89in)

WAR-TIME SERVICE?
Controversy has raged over whether the IS-3 saw action in the war's last days or not. During the Cold War it was believed that the first 29 IS-3 tanks completed were rushed to Berlin and joined the last phase of the struggle to capture the German capital. However, after the end of the Cold War, research at the newly opened Russian archives suggested that while the tanks did arrive, they did not actually participate in any active combat.

SELF-PROPELLED GUNS

The Red Army augmented its tank fleet's combat power by employing self-propelled guns (SPGs), which could easily be mass-produced. Typically an SPG mounted a powerful gun in a fixed armoured superstructure atop a tank chassis.

The following self-propelled guns are included in this chapter:

- ZiS-30 Tank Destroyer
- KV-2 Self-propelled 152mm Howitzer
- SU-76 Assault Gun
- SU-85 Self-propelled Gun
- SU-100 Self-propelled Gun
- SU-122 Self-propelled Gun
- SU-152 Heavy Self-propelled Gun
- ISU-122 Heavy Self-propelled Gun
- ISU-152 Heavy Self-propelled Gun

A battery of ISU-152 assault guns of the Fourth Tank Army returns from the front for resupply and repainting in the spring of 1944.

ZiS-30 Tank Destroyer

During the first months of Operation 'Barbarossa', the Soviets suffered catastrophic tank casualty rates. To swiftly replace these losses they resorted to desperate expedients such as the ZiS-30 tank destroyer.

As the Axis Panzer spearheads charged inexorably eastward during the summer of 1941 they decimated the Red tank force. The Soviets now needed to get as many armoured vehicles to the front line as quickly as possible. This led to desperate expedients that used existing components to quickly form improvised fighting vehicles. Swiftly-designed during July–August 1941, the ZiS-30 light tank destroyer used as its chassis the existing Komsomolyets artillery tractor, of which over 4000 had been produced in the late 1930s. The Komsomolyets was a fully-tracked lightly armoured prime-mover that mounted a 7.62mm (0.3in) DT machine-gun in its hull front.

In late summer 1941 Factory #92 at Gorky simply fitted the potent and immediately available 57mm (2.24in) ZiS-29 anti-tank gun in an open mounting on the Komsomolyets tractor. During September and October Factory #92 utilized the available tractor chassis to produce 101 ZiS-30 vehicles.

The self-propelled anti-tank gun's expedient nature limited its tactical effectiveness; it was unstable and overloaded, the crew lacked adequate protection and only limited rounds could be stowed. By July 1942 all the ZiS-30s produced had either been lost in combat or abandoned.

KhTZ

The two-man KhTZ (or KhTZ-16 tractor tank) was another expedient Soviet light tank destroyer swiftly thrown together during autumn 1941's desperate days. KhTZ's Kharkov factory took 61 available KhTZ-3 artillery tractors and fitted them with a simple fixed armoured hull superstructure. Inside the superstructure's raised rear section were mounted a 45mm (1.77in) anti-tank gun and a co-axial 7.62mm (0.3in) DT machine-gun.

ZiS-30

This was an evidently expedient design that mounted the large 57mm (2.24in) ZiS-29 gun on the light Komsomolyets artillery tractor.

ZiS-30 tank destroyer

Crew 4
Production 1941
Weight 4 tonnes (4.4 tons)
Dimensions 3.45m (11ft 4in) x 1.86m (6ft 1in) x 2m (6ft 6in)
Engine 37kW (50hp) GAZ-M 4-cylinder
Road Speed 40km/h (25mph)
Range 250km (155 miles)
Armament 1 x 57mm (2.24in) ZiS-2 gun; 1 x 7.62mm (0.3in) DT machine-gun
Armour 7–10mm (0.28–0.39in)

KhTZ tank destroyer

Obvious here is the distinctive paired bogie-wheel and track arrangement of the KhTZ-3 tractor and the crudely-fashioned improvised hull superstructure.

KV-2 Self-propelled 152mm Howitzer

The enormously tall box-like turret of the KV-2 made this behemoth one of the most distinctive Soviet armoured vehicles deployed in the bitter struggle for survival that raged across the Eastern Front during 1941–45.

KV-2 Model 1939

Weighing 12.9-tonnes (14.2 tons), the extremely tall and large sloped-fronted cuboid MT-1 turret identifies this vehicle as a KV-2 Model 1939.

KV-2 Model 1939
Crew 5
Production 1939–40
Weight 53.6 tonnes (48.1 tons)
Dimensions 7.31m (23ft 11in) x 3.49m (11ft 5in) x 3.93m (12ft 1in)
Engine 373kW (500hp) V-2 diesel
Road Speed 25km/h (12mph)
Range 200km (120 miles)
Armament 1 x 152 mm (5.98in) 1938/1940 L20 howitzer; 1 x co-axial 7.62mm (0.3in) DT machine-gun
Armour 30–110mm (1.18–4.3in)

During 1938–40, alongside the KV-1 heavy tank, Soviet engineers also designed the KV-2 heavy artillery tank (self-propelled howitzer). The KV-2 comprised the KV-1's chassis and shallow-hull superstructure together with a large newly-designed turret. The five-man KV-2 Model 1939 had the distinctive extremely large, very tall frontally-sloped heavily-armoured cuboid MT-1 turret. The later version, the six-man Model 1940, mounted a large but not as tall vertically-fronted box-like MT-2 turret with a 7.62mm (0.3in) DT machine-gun in the rear.

Both turrets mounted the short-barrelled 152mm (5.98in) Model 1938/40 L/20 heavy howitzer in a large box-like mantlet. Tactically, the KV-2 was primarily employed as a 'bunker-busting' armoured assault or breakthrough vehicle – firing a special anti-concrete round it could penetrate even thickly-protected enemy fortifications. The sheer height and weight of the turret and gun rendered the KV-2 design top-heavy and unstable, particularly when moving across undulating ground. The heavily-protected KV-2 featured 30–110mm- (1.18–3.15in-) thick armoured plates.

The KV-2's suspension/track arrangement consisted of six medium-sized solid-disc road-wheels on each side, each connected to a torsion-bar suspension, with a large-sized rear driving wheel and a large front idler.

KV-2 Model 1940

The vertically-faced slightly less tall MT-2 turret design identifies this as the KV-2 Model 1940 variant.

Later models

Unsurprisingly, the KV-2 Models 1939 and 1940 respectively weighed a colossal 53.9 tonnes (59.4 tons) and 57.9 tonnes (63.8 tons). These later models also featured the 152mm (5.98in) M-10T howitzer. The Model 1940 also included a rear turret DT machine-gun.

Between late 1939 and mid-1941 Soviet factories produced 334 KV-2 vehicles.

SU-76 Assault Gun

Manufactured in two basic designs, the SU-76 light assault gun became, after the famous T-34 medium tank, the most produced Soviet armoured fighting vehicles of World War II.

By late 1941 the Soviets had recognized that they urgently needed to produce as many armoured vehicles as possible with the minimum investment of time and resources. SPGs – 'turretless tanks' – were an obvious way of producing numerous armoured vehicles quickly and cheaply.

During 1942, therefore, Factory #38 in Kirov and Factory #92 at Gorky worked together to design the SU-76 assault gun. The vehicle was primarily intended to fulfil the tactical role of infantry support, but its main armament had some armour-piercing capability, so the vehicle could also function as a light tank destroyer.

The SU-76's chassis was a modified lengthened and widened version of the T-70 light tank chassis. The SU-76's 'wedged' track/suspension arrangement, however, had an extra (sixth) large non-bogie doubled-tyred road-wheel added to support the vehicle's increased weight. The vehicle also retained the large front track driving-wheel, the large rear idler wheel and the three relatively large top-track return rollers featured in the original T-70 tank.

The interior design of the SU-76 was, however, profoundly different from that of the T-70. The engine, fuel tanks and driver's station were at the front of the vehicle, permitting the construction of a spacious lightly armoured closed-topped fighting compartment that covered the rear half of the vehicle. In the centre-front of this fighting compartment the SU-76 mounted the large and potent long-barrelled 76.2mm (3in) ZiS-3Sh L/42.6 variant of the ZiS-3 L/41 divisional cannon. Perhaps the most externally distinctive feature of the SU-76 was the extremely large and very long gun mantlet that projected well forward of the fighting compartment's frontal face.

SU-76

The SU-76 is differentiated externally from the succeeding SU-76M by its roofed-over fighting compartment that featured steeply-sloped rear plates.

SU-76 (SU-12)

Crew 3
Production 1942–43
Weight 10.3 tonnes (11.4 tons)
Dimensions 5m (16ft 5in) x 2.7m (8ft 10in) x 2.1m (6ft 11in)
Engine 2 x mated (in parallel) 52kW (72hp) GAZ-202 6-cylinder petrol
Road Speed 45km/h (28mph)
Range 354km (240 miles)
Armament 1 x 76.2mm (3in) ZiS-3Sh L/42.6 gun
Armour 16–35mm (0.59–1.4in)

The SU-76 was powered by the same paired 52kW (70hp) GAZ-202 engines as in the T-70 tank. Although the SU-76 could obtain a respectable maximum road speed of 45km/h (28mph) and a peak operational range

ENGINE PROBLEMS

The same mated pair of GAZ-202 engines used in the T-70 also powered the SU-76. In the latter, however, these engines were reconfigured 'in parallel' in the vehicle front, rather than mated one behind the other as in the T-70; each engine controlled one track. This arrangement proved most unsatisfactory: it was challenging for the driver to control both engines simultaneously and excessive vibration led to key components requiring regular workshop repairs.

of 354km (220 miles), the vehicle's tactical mobility and engine reliability were very poor (see box 'Engine Problems'). These issues led the Soviets to suspend SU-76 production in March 1943. During a four-month production run that commenced in December 1942, Soviet factories manufactured just 320 SU-76 vehicles.

SU-76M

The early 1943 production halt enabled the Soviets to resolve the engine problems that had bedevilled the SU-76 assault gun. The designers reconfigured the chassis so that the paired 52kW (70hp) GAZ-202 engines were positioned one behind the other along the vehicle's right-hand side. This new power-plant configuration proved far superior than the previous one. In June 1943 the solving of this key deficiency enabled the Soviets to commence large-scale mass production of the modified vehicle – which was now designated the SU-76M.

At the same time as the engine re-configuration, the armoured roof was removed from the vehicle. Although unpopular with the four-man crew, this step facilitated better access to the main armament and thus made servicing and maintaining the weapon easier than it had been previously.

The shape of the fighting compartment screen was also altered with its rear side armoured plates sloping down at a 45-degree angle to leave only a low protective screen across the vehicle's rear. This reduced the amount of armoured plate required to complete an SU-76M, which facilitated cheaper mass-production.

With the key deficiencies of the SU-76 eradicated, the mighty Soviet tank production machine was increasingly geared-up to churn out large numbers of SU-76M assault guns quickly and cheaply. By the time the SU-76M was reaching the front lines in large numbers during late 1943, however, the lethality of its 76.2mm (3in) ZiS-3 gun had been eroded due to increases in the armour possessed by enemy tanks.

Despite this, the SU-76M could not only still provide excellent tactical service in the infantry support role (which it monopolized for the rest of the

SU-76M

Crew 4
Production 1943–45
Weight 10.2 tonnes (11.2 tons)
Dimensions 5m (16ft 5in) x 2.7m (8ft 10in) x 2.1m (6ft 11in)
Engine 2 x mated (in-line) 52kW (72hp) GAZ-202 6-cylinder petrol
Road Speed 45km/h (28mph)
Range 320km (199 miles)
Armament 1 x 76.2mm (3in) ZiS-3Sh L/42.6 gun
Armour 16–35mm (0.59–1.4in)

SU-76M (three-view)

This view of the SU-76M clearly reveals its open-topped fighting compartment that featured its side plates' rear portion sloped down at a 45-degree angle.

SU-76i
Crew 4
Production 1943
Weight 22 tonnes (24.3 tons)
Dimensions 5.41m (17ft 9in) x 2.95m (9ft 8in) x 2.38m (7ft 9in)
Engine 220kW (300hp) Maybach HL120-TRM V-12
Road Speed 40km/h (25mph)
Range 180km (112 miles)
Armament 1 x 76.2mm (3in) S-1 gun
Armour 15–35mm (0.59–1.38in)

war), but also had a satisfactory indirect fire capability. While never popular with its crews, who felt exposed in the open-topped fighting compartment, the SU-76M could be mass-produced quickly and cheaply. Between July 1943 and June 1945 three Soviet factories produced an impressive 13,932 SU-76M assault guns.

SU-76i

During the early 1943 hiatus between SU-76 and SU-76M mass production, the Soviets investigated producing a stop-gap alternative. The Red Army had captured 201 German Panzer III tanks and tests confirmed that this might be a satisfactory platform for mounting the 76.2mm (3in) ZiS-3Sh L/42.6 gun fitted in the SU-76. The Soviet Factory #37 at Sverdlovsk quickly designed and converted these Panzer IIIs into an improvised assault gun, designated the SU-76i – the 'i' meant 'foreign'. This design fitted a simple sloped-sided armoured box superstructure onto the frontal portion of the Panzer III's hull decking. This superstructure had 15–35mm (0.59–1.38in) thick armoured plates with the hull being protected by 20–30mm (0.78–1.18in) armour.

Although the SU-76i was designed to mount the ZiS-3Sh gun, some technical complications instead led to the mounting of the similar 76.2mm (3in) S-1 gun. Given its placement in

SU-76i
As a stop-gap the Soviets fitted an armoured superstructure to captured German Panzer III tanks to create the SU-76i assault gun.

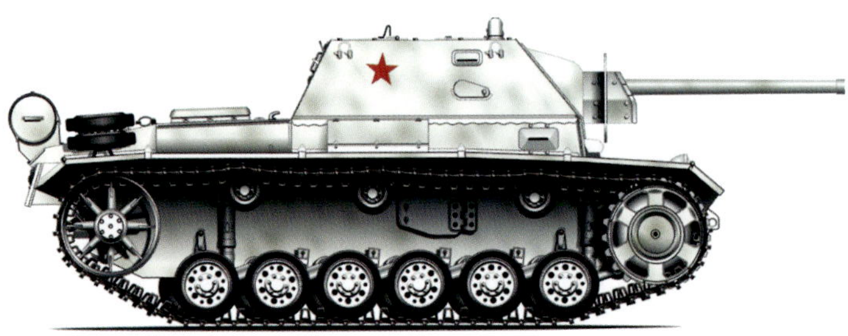

SU-76i (Winter camouflage)
This SU-76i assault gun displays a typical winter camouflage scheme of a whitewash coat roughly added to the normal olive green finish.

SU-76i VARIANTS

In addition to the available captured German Panzer III tank chassis, the Soviets also fitted the standard SU-76i style assault gun armoured superstructure onto other available captured German chassis. At least eight StuG III assault guns, for example, were converted in this fashion. The slightly different dimensions of the German chassis employed, of course, necessitated slightly different configurations for the slope-sided armoured box superstructure fitted upon the hull decking.

the forward-located superstructure's front face, it therefore overhung the vehicle's front. This gun when mounted in the SU-76i featured a much smaller mantlet than the SU-76 or SU-76M; this mantlet resembled that of the KV-1 Model 1940 tank.

The SU-76i retained the Panzer III's original power plant, the 220kW (300hp) Maybach HL120-TRM V-12 engine, and could obtain a top by-road speed of 40km/h (25mph).

The mobility and power of the SU-76M proved very useful in an infantry support role in the confined streets of Berlin.

SU-85 Self-propelled Gun

To rush the potent 85mm (3.35in) gun into combat, the Soviets designed an easy-to-manufacture tank destroyer – the SU-85 – that took many of its components from the existing T-34 and SU-122 designs.

Su-85

The white eagle insignia on this SU-85's superstructure identifies it as belonging to a unit of the Soviet-raised People's Army of Poland (LWP).

During December 1942 the Soviet forces deployed on the Eastern Front's northern sector first encountered the new German Panzer VI Tiger heavy tank. The combat experiences thus gained led the Soviets to doubt the future viability of the 76.2mm (3in) gun as an effective tank killer against such well-protected German tanks. When these doubts first surfaced the Soviets began design work on a new assault gun or tank destroyer – the SU-85 – that would mount the potent 85mm (3.35in) D-5T gun intended for the future T34/85 medium tank, which was then in an early development phase.

The SU-85's mission was to engage and knock out German tanks including the formidable new Tiger. The SU-85 also had to be easy to mass produce in large numbers using existing components. Thus this new Soviet tank destroyer was based to a significant degree on the existing SU-122 self-propelled howitzer design, which itself shared four-fifths of its component parts with the proven T-34 medium tank.

SU-85M
The last 315 production vehicles in the SU-85 manufacturing run, produced during late 1944, featured several modifications; these vehicles were re-designated as the SU-85M, with the 'M' signifying 'modified' in the Russian language. The main innovation was that these vehicles utilized the up-armoured chassis of the new SU-100 tank destroyer. These SU-85M vehicles also featured a re-designed gun mantlet as well as a modified commander's cupola on the superstructure roof.

Thus the SU-85 had just one-quarter of its components as new designs, with the remaining parts being taken directly from either the T-34 or SU-122 designs.

Production begins

The SU-85's design work unfolded during spring 1943. Subsequently, during the summer of 1943 UZTM's 'Uralmash' factory in Sverdlovsk began mass production of the four-man SU-85.

The new vehicle mounted an 85mm (3.35in) D-5S gun, a modification of the D-5T destined for the T34/85 that had been designed specifically for use in 'turretless tanks'. The gun was mounted, offset to the right, in a large mantlet fitted on the front of the shallow sloped-sided armoured superstructure. The latter was positioned well forward on the SU-85's chassis. This arrangement gave the vehicle a low silhouette as well as a very significant barrel overhang, amounting to some 45 per cent of its hull length. The SU-85 featured no machine-guns as secondary armament and carried 60 rounds for its 85mm (3.35in) gun.

Weighing 29.6 tonnes (32.6 tons), the SU-85 was protected by 30mm- (1.18in-) thick well-sloped side and rear armoured plates. On the sloped frontal plate the tank destroyer featured even thicker 75mm (2.95in) armoured plates. In addition, at just 2.45m (8ft) high the vehicle had a low silhouette and this aided its survivability by decreasing the amount of target at which the enemy could fire.

Performance

Powered by a 373kW (500hp) rated V-2 diesel engine, and with the T-34's proven Christie vertical coil-spring suspension, the SU-85 could obtain an impressive top road speed of 55km/h (34mph). Completed vehicles were typically fitted with rubber-rimmed road-wheels, but some featured one or more metal-rimmed wheels due to continuing rubber shortages throughout the war period. The vehicle could obtain a maximum operational range during road travel of 400km (250 miles). This was made possible by its internal 540 litre (119 gal) fuel tanks, which was later augmented by the fitting of four cylindrical external fuel tanks mounted on the sloped sides of the hull superstructure's rear section.

With the entry of the T-34/85 and SU-100 into mass-production from December 1943 and late 1944, respectively, the need for the SU-85 soon diminished. Therefore production of the SU-85 ceased in December 1944 after 2050 vehicles had been completed.

SU-85
Crew 4
Production 1943–44
Weight 29.6 tonnes (32.6 tons)
Dimensions 8.15m (26ft 8in) x 3m (9ft 10in) x 2.45m (8ft)
Engine 373kW (500hp) V-2 12-cylinder diesel
Road Speed 55km/h (34mph)
Range 400km (249 miles)
Armament 1 x 85mm (3.35in) D-5T L/51.6 gun
Armour 20–45mm (0.78–1.77in)

SU-85
This SU-85M from a LWP unit features a re-designed gun mantlet and commander's cupola, as well as an un-armoured chassis akin to the SU-100.

SU-100 Self-propelled Gun

The Soviets produced the potent SU-100 tank destroyer to supersede the SU-85 by marrying an up-armoured modification of the latter's hull and chassis to the newly-developed 100mm (3.94in) D-10S cannon.

The late 1943 introduction of the T34/85 tank, fitted with the same 85mm (3.35in) weapon as the SU-85, brought the its future role into question. Consequently, the Soviets produced the SU-100 tank destroyer. This low-silhouetted 31.6 tonne (34.8 ton) vehicle married a potent 100mm (3.94in) gun to an up-armoured SU-85 chassis. The four-man SU-100 mounted the new long-barrelled 100mm (3.94in) D-10S Model 1944 cannon. This was fitted, offset to the right, in a large rounded mantlet attached to the superstructure's well-sloped front plate. This plate was located well forward in the vehicle; given the D-10S cannon's length, the vehicle's barrel overhang amounted to 59 per cent of the hull's entire length. Like its predecessor, the SU-100 had no machine-gun fitted for close-range defence. It also mounted a 10-RF-26 radio set.

Up-armoured SU-85

The SU-100 hull was an up-armoured modified SU-85 chassis (itself based on that of the T-34). It was protected by 20–110mm- (0.78–4.33in-) thick armour. Powered by a 373kW (500hp) rated V-2 12-cylinder diesel engine, the SU-100 could reach a top road speed of 35km/h (22mph) and a maximum by-road operating range of 300km (186 miles). During late 1944 and early 1945 the Soviets produced 1534 SU-100 tank destroyers, many of which participated in the final Soviet advance into central Germany.

SU-100
Crew 4
Production 1944–45
Weight 31.6 tonnes (34.8 tons)
Dimensions 9.45m (31ft) x 3m (9ft 10in) x 2.25m (7ft 5in)
Engine 388kW (520hp) V-2-34M 12-cylinder diesel
Road Speed 48km/h (30mph)
Range 320km (199 miles)
Armament 1 x 100mm (3.94in) D-10S L/53.5 gun
Armour 20–75mm (0.78in–2.95in)

SU-100
Painted with a distinctive white horizontal identification stripe, this First Guard's Tank Army SU-100 participated in the climactic April 1945 battle of Berlin.

SU-100
Compare the SU-100's longer-barrelled cannon with that of the SU-85M (opposite); both designs sported the re-designed commander's cupola.

SU-122 Self-propelled Gun

The Soviets employed the proven chassis of the T-34 tank as a platform to mount the powerful 122mm (4.8in) field howitzer in an armoured superstructure to create the SU-122 SPG.

The five-man SU-122 was one of the earliest Soviet self-propelled guns. During April 1942 Soviet design teams at UZTM and Factory #592 began development work on a medium self-propelled gun, initially designated the U-35. To simplify production, the vehicle was based on the existing chassis of the combat-proven T-34 medium tank. The SU-122 would mount the potent 122mm (4.8in) M-30 Model 1938 L/23 field howitzer, offset to the right, in an armoured sloped superstructure atop the T-34 chassis.

The gun was fitted to a large and long box-like gun mantlet, which was mounted low down in the superstructure's well-sloped frontal plate. Like other Soviet SPGs, the SU-122 had no machine-guns.

The SU-122 first entered production in December 1942 and reached frontline units from early 1943. The vehicle's primary tactical role was to provide direct-fire support for infantry attacking prepared enemy defences. From the summer of 1943, moreover, the SU-122's mediocre anti-armour capability was marginally improved, thanks to the introduction of a new HEAT (high explosive anti-tank) round for the M30 gun.

Weighing 30.5 tonnes (33.6 tons), the SU-122 was protected by well-sloped 20–45mm (0.79–1.77in) thick armour. The provision of a large fighting compartment, fitted with a 9R radio set, enabled an enlarged crew of five to service the gun's fire with relative ease. Powered by a 373kW (500hp) V-2 diesel engine, the vehicle could obtain a top road speed of 55km/h (34mph).

In addition to its internal fuel tanks, the SU-122 was fitted with additional cylindrical external fuel tanks attached to the rear hull. These enabled the vehicle to attain a maximum by-road operational range of 300km (186 miles).

During January–November 1943 Soviet factories produced 1150 SU-122 vehicles. These served in Soviet medium SP artillery regiments that consisted of 16 SU-122s plus a T-34 command tank.

SU-122

These two views illustrate the SU-122's well-sloped low silhouette and the large gun mantlet similar to that of the SU-76.

SU-122
Crew 5
Production 1943
Weight 30.5 tonnes (33.6 tons)
Dimensions 6.95m (22ft 10in) x 3m (9ft 10in) x 2.32m (7ft 7in)
Engine 373kW (500hp) V-2 diesel
Road Speed 55km/h (34mph)
Range 300km (186 miles)
Armament 1 x 122mm (4.8in) M-30 Model 1938 L/23 field howitzer
Armour 20–45mm (0.79–1.77in)

SU-122 (winter camouflage)

This mid-production SU-122 served in Volkhov during the winter of 1943. It features improvised camouflage over washable white paint.

SU-152 Heavy Self-propelled Gun

The last of the KV-1 family of vehicles, the mighty SU-152 assault gun mounted the colossal 152.4mm (5.98in) gun in a well-protected superstructure fitted to the chassis of the KV-1 tank.

SU-152

This vehicle featured a monotone drab olive paint scheme while on the Karelian Front in the summer of 1944.

The valuable experience gained during the December 1942 encounters with the new German Tiger tank near Leningrad prompted the Soviets to swiftly design a heavier, more lethal and better-protected prototype KV-14 assault gun in January 1943. In February 1943 this prototype entered mass production as the SU-152 assault gun. The five-man SU-152 mounted the lethal 152.4mm (5.98in) ML-20S Model 1938 L/29 field gun-howitzer, offset to the right, in a heavily armoured, well-sloped superstructure. This forward-positioned casement-style fighting compartment was constructed on top of the proven chassis of the KV-1 heavy tank.

SUSPENSION SPECIFICATIONS

The SU-152 assault gun featured a suspension/track arrangement identical to that of the KV-1. This consisted of six medium-sized solid-disc road-wheels per side, each independently connected to a torsion-bar suspension, together with a rear driving wheel and a front idler. In addition, the track top ran over three high-positioned small-sized return rollers. The vehicle also had the wide tracks of the KV-1, which helped it maintain reasonable tactical mobility despite its considerable weight.

Potent gun

Firing Armour-Piercing (AP) rounds, the ML-20S gun had a very potent lethality under test conditions. At the extreme combat range of 2000m (6560ft) the ML-20S gun could penetrate 110mm (4.3in) of vertical armour. The likelihood of the vehicle achieving a hit at this great range, however, was low due to the provision of mediocre optics and the tactical limitations of the crew. That said the SU-152 gained a reputation within the Red Army for its respectable lethality against German tanks, and soon gained the nickname 'Beast Slayer'.

Despite this, the SU-152's primary tactical role remained infantry support, where it typically fired high explosive (HE) rounds. Some SU-152s repaired at ChKZ received a 12.7mm (0.5in) DShK heavy anti-aircraft machine-gun mounted on the superstructure roof.

SU-152
Crew 5
Production 1943
Weight 45.5 tonnes (50.2 tons)
Dimensions 8.95m (29ft 4in) x 3.25m (10ft 8in) x 2.45m (8ft)
Engine 473kW (600hp) V-2 12-cylinder diesel
Road Speed 43km/h (27mph)
Range 220km (137 miles)
Armament 1 x 152.4mm (5.98in) ML-20S Model 1938 L/29 gun-howitzer; sometimes 1 x 12.7mm (0.5in) DShK heavy machine-gun
Armour 20–75mm (0.78–2.95in)

The vehicle featured either a 9R, 10R or 10RK-26 radio device.

Weighing a mighty 45.5 tonnes (50.2 tons), the SU-152 was protected, respectively, by 75mm (2.95in) front, 60mm (2.36in) side/rear and 20mm (0.78in) roof armour plates. Featuring a suspension and track arrangement that was identical to that of the KV-1 tank, the SU-152 was powered by a 473kW (600hp) up-rated V-2 diesel engine. This, in combination with wide tracks, enabled the colossal SU-152 to still obtain a respectable maximum road speed of 43km/h (27mph), although its top cross-country speed was only half this. The vehicle was fitted with rear-mounted cylindrical external fuel tanks that boosted its fuel capacity to a considerable 945 litres (208 gal); this fuel load enabled the 'gas-guzzler' to obtain a still decent operational range of 220km (137 miles). During February–December 1943 the Soviets produced 704 SU-152s.

SU-152

The stretching of the upper track over the three high-positioned return rollers is clearly evident in this side view of an SU-152.

ISU-122 Heavy Self-propelled Gun

Marrying powerful guns with 'turretless tanks' based on the IS heavy tank chassis, the Soviets created the ISU family of multi-role assault guns. The ISU-122 wielded the potent 122mm (4.8in) gun.

ISU-122

This ISU-122 has been finished in two-tone camouflage and took part in the assault on Berlin in April 1945.

ISU-122

Crew 4 or 5
Production 1944–45
Weight 45.5 tonnes (50.2 tons)
Dimensions 9.85m (32ft 4in) x 3.07m (10ft 1in) x 2.84m (8ft 2in)
Engine 388kW (520hp) V-2-IS diesel
Road Speed 37km/h (23mph)
Range 220km (137 miles)
Armament 1 x 122mm (4.8in) A-19S Model 1931/37 L/46.5 gun; sometimes 1 x 12.7mm (0.5in) DShK heavy machine-gun
Armour 30–120mm (1.18–4.7in)

Once IS heavy tanks entered service Soviet designers began developing derivative heavy assault guns. Initially, these designers developed the ISU-152 based on the IS chassis that mounted the 152.4mm (5.98in) gun. However, there were delays in this weapon's manufacture, so during December 1943 ChKZ modified the design to produce the ISU-122. This was essentially an ISU-152 with the main armament exchanged for a 122mm (4.8in) A-19S gun, which also sported modified internal round stowage. It was manned by a crew of four (or five if including a second loader).

The ISU-122 was a low-silhouetted assault gun that mounted the large gun in a massive box-like mantlet identical to that of the ISU-152. The cannon was

ISU-122 heavy assault guns move into the suburbs of Berlin in April 1945.

offset to the right in the well-sloped front of the casement-style fighting compartment, which was located well forward. The mantlet's low-slung 'snout-tip' reached the very front of the vehicle, meaning that the entire barrel length overhung the vehicle hull. From October 1943 onwards, vehicles sent back for factory repairs were outfitted with the 12.7mm (0.5in) DShK heavy machine-gun mounted on the fighting compartment's roof.

Weighing 45.5 tonnes (50.2 tons), the ISU-122 was protected by a 120mm (4.7in) mantlet, 90mm (3.54in) front, 60mm (2.36in) side/rear and 30mm (1.18in) roof armour. The vehicle sported the IS-style torsion-bar suspension and running gear with six independent solid-disk road-wheels per side and three high-positioned return rollers. Powered by an a 388kW (520hp) V-2-IS diesel engine, the ISU-122 could attain a maximum road speed of 37km/h (23mph) and a by-road range of 220km (137 miles).

From August 1944 the modified ISU-122S appeared. This mounted the longer-barrelled 122mm (4.8in) D-25S gun fitted in a re-designed mantlet that lacked the low-slung 'snout'. This gun had an improved rate of fire and reduced recoil in comparison with the A-19S gun. Between March 1944 and November 1945, ChKZ produced 1735 ISU-122 and 675 ISU-122S vehicles.

ISU-122S

Crew 4 or 5
Production 1944–45
Weight 45.53 tonnes (50.2 tons)
Dimensions 9.85m (33ft 11in) x 3.07m (10ft 1in) x 2.84m (8ft 2in)
Engine 388kW (520hp) V-2-IS diesel
Road Speed 37km/h (23mph)
Range 220km (137 miles)
Armament 1 x 122mm (4.8in) D-25S gun; sometimes 1 x 12.7mm (0.5in) DShK heavy machine-gun
Armour 30–120mm (1.18–4.7in)

ISU-122S

The modified ISU-122S can be differentiated from its predecessor due to the longer gun with muzzle brake and the 'snoutless' gun mantlet.

ISU-152 Heavy Self-propelled Gun

The Soviets fitted the mighty 152.4mm (5.98in) gun to a heavily-protected casement-style armoured superstructure fitted on top of the IS heavy tank chassis to create the renowned ISU-152 multi-role SPG.

ISU-152

This late-production ISU-152 incorporates a 12.7mm (0.5in) DShK anti-aircraft machine-gun mounted on its fighting compartment roof.

The first heavy assault gun variant based on the chassis of the IS heavy tanks that the Soviets developed was the ISU-152. The development work carried out by ChKZ during late 1943 culminated when the ISU-152 Model 1943 entered mass production in December. Externally there were only modest differences in the appearance of the SU-152 and the ISU-152. The latter had a slightly taller and shorter armoured fighting compartment, as well as a large gun mantlet with a more pronounced low-slung 'snout'. There were also minor differences in the chassis of the two vehicles – the former being based on that of the KV-1 and the latter on that of the IS heavy tank family.

The ISU-152 mounted the same potent 152.4mm (5.98in) ML-20S Model 1938 L/29 gun as carried by the SU-152. Like all Soviet SPGs, this weapon was set in a large gun mantlet that was offset to the right of the vehicle's centre-line. The gun mantlet sat positioned in the middle height-wise of the vehicle's heavily armoured, well-sloped superstructure frontal plate. The increased height of the casement-style fighting compartment gave the crew rather more internal working space, which enabled them to operate the gun more effectively in tactical situations. Given their weight the vehicle carried just 18 main armament rounds, which proved to be a liability during combat.

ISU-152

Crew 4 or 5
Production 1943–44
Weight 47.3 tonnes (52.1 tons)
Dimensions 9.18m (30ft 1in) x 3.07m (10ft 1in) x 2.48m (8ft 2in)
Engine 388kW (520hp) up-rated V-2-IS
Road Speed 40km/h (25mph)
Range 220km (137 miles)
Armament 1 x 152.4mm (5.98in) ML-20S Model 1938 L/29 gun
Armour 30–120mm (1.18–4.72in)

ISU-152 (wnter camouflage)

This ISU-152, finished off in winter whitewash camouflage, has painted along its barrel the familiar slogan of 'Forward to the destruction of the enemy'.

Red Army soldiers ride on the warm rear deck of an ISU-152 in Poland, January 1945.

Tactical roles

The ISU-152 performed three main tactical roles: firing various Armour-Piercing (AP) rounds, the vehicle acted as a tank destroyer endeavouring to knock out enemy AFVs; providing direct fire with high-explosive (HE) rounds, the ISU-152 was used for infantry support; finally – again delivering HE via indirect fire – it performed as self-propelled artillery.

The ISU-152 weighed a considerable 47.3 tonnes (52.1 tons), slightly more than the SU-152. In part, this was due to the relatively high levels of protection that the vehicle incorporated. The ISU-152 featured 30mm (1.18in) roof armour plates and 75–90mm (2.95–3.54in) side plates. Its frontal protection was 60–90mm (2.36–3.54in) thick, while the armour on the mantlet was 120mm (4.72in) thick. Like other Soviet self-propelled guns, the vehicle's survivability was enhanced by the small target it presented – the low-silhouetted ISU-152 was just 2.48m (8ft 2in) tall.

The four-man vehicle was powered by a 388kW (520hp) up-rated V-2-IS and its IS-based chassis featured six road-wheels per side, independently connected to torsion bars.

Despite its size and weight, the ISU-152 obtained relatively decent tactical mobility, with a top by-road speed of 40km/h (25mph). In late 1944 the improved ISU-152 Model 1944 was introduced. This modified variant featured the improved ML-20SM gun, significantly increased mantlet armour and a 12.7mm (0.5in) DShK heavy machine-gun mounted on the hull roof.

Between December 1943 and autumn 1945 ChKZ produced 1885 ISU-152 SPGs.

ISU-152 (summer camouflage)

This ISU-152 had a four-tone tactical camouflage scheme applied to it during the July–August 1944 Lvov–Sandomierz Offensive.

LEND-LEASE TANKS

During 1941–45, the Western Allies dispatched thousands of 'Lend Lease' tanks to augment the combat power of the beleaguered Red Army; these vehicles contributed significantly to the eventual Soviet defeat of the Axis invaders.

This chapter includes the following tanks:

- Infantry Tank Mk.II Matilda
- Infantry Tank Mk.III Valentine
- Infantry Tank Mk.IV Churchill
- Light Tank Mk.VII Tetrarch
- Light Tank M3 Stuart
- M3 General Lee
- M4A2 Sherman
- M10 Wolverine

An M4A2 Sherman of the I Guards Mechanised Corps, 3rd Ukrainian Front, advances during fighting in Austria, April 1945.

Infantry Tank Mk.II Matilda

The slow but heavily-armoured British Matilda Mk.II infantry support tank provided sterling service to the hard-pressed Soviet forces who were locked into a desperate defence of Moscow during the winter of 1941–42.

During the period October 1941 through to March 1943, the British dispatched 1084 examples of the Infantry Tank Mark II (A12) Matilda to the Red Army via the north Atlantic convoy route, of which 918 arrived. The vehicles employed by the Soviets spanned several variants, including the Matilda II, IIA, III and IV. Weighing 26.92 tonnes (29.7 tons), the four-man Matilda II was a heavily-protected infantry support tank.

Above: Matilda IIA

Although both variants were externally almost identical, in the Mark IIA a co-axial BESA machine-gun replaced the Vickers of the Matilda II. This tank fought with the 133rd Tank Brigade in May 1942.

Main gun

The tank mounted a long-barrelled '2-pounder' 40mm (1.57in) OQF L/52 gun in a wide half-cylindrical mantlet on the front of its cast rounded-edge cuboid turret. The latter was positioned reasonably well forward on the hull along the central axis. By way of secondary armament, the tank featured a co-axial 7.7mm (0.303in) Vickers machine-gun. The Matilda II was heavily protected: its armoured plates varied in thickness between 20mm (0.78in) on the decking and 78mm (3.1in) on the glacis plate.

The heavy Matilda II remained relatively under-powered given that its power-plant consisted of two mated 6-cylinder 65kW (87hp) AEC petrol engines. Consequently, the Matilda II could only obtain an unimpressive maximum road speed of 24km/h (15mph), and just 10km/h (6mph) cross-country. The tank could attain a maximum operational range when travelling on roads of 257km (160 miles). The Matilda's track and suspension arrangement comprised five pairs of bogie-mounted small-sized road-wheels per side connected to horizontal coil springs, and a large driver and idler wheel per side. The tracks were covered by distinctive extensive armoured skirting plates. The Soviets also received the later Matilda IIA, III and IV variants, which incorporated minor modifications; the Mark III and IV, for example, had up-rated Leyland engines.

Matilda IV

This up-powered Mark IV of the 38th Tank Brigade is adorned with foliage camouflage.

Matilda IIA

Crew 4
Production 1940–1942
Weight 26.92 tonnes (29.7 tons)
Dimensions 5.6m (18ft 5in) x 2.59m (8ft 6in) x 2.5m (8ft 3in)
Engine 2 x mated 6-cylinder 65kW (87hp) AEC petrol
Road Speed 24km/h (15mph)
Range 257km (160 miles)
Armament '2-pounder' 40mm (1.57in) OQF L/52 gun; 1 x co-axial 7.7mm (0.303in) BESA machine-gun
Armour 20–78mm (0.78–3.1in)

Infantry Tank Mk.III Valentine

The Valentine was the most numerous infantry support tank the Western Allies sent to assist the Red Army; this simply-engineered and robust vehicle made a significant contribution to the Soviet defensive stands and counter-attacks mounted during 1942–43.

Valentine Mk.II

This Soviet Valentine II lacks a turret left-side pistol port; the turret rivet-heads are also clearly visible.

The Western Allies dispatched 3784 Infantry Tank Mark III Valentine vehicles to the Soviet Union during 1941–43. The Red Army employed many different Valentine variants, including the Mark II, IV, V, VI, IX and X tanks.

The three-man Valentine was a medium infantry support tank. At 17.3 tonnes (19.1 tons), the Valentine II was lighter than the Matilda II. This was largely due to its inferior protection; the tank's armoured plates ranged in thickness from just 8mm (0.32in) on the decking through to 65mm (2.56in) on the turret front.

Valentine Mk VII

A Fifth Guards Tank Brigade Valentine Mk VII, Caucasus, early 1943. The unit can be identified by the white diamond on the side of the turret.

The Valentine mounted as its main armament a long-barrelled '2-pounder' 40mm (1.57in) OQF L/52 gun in a narrow rounded mantlet on the front of its curved medium-sized riveted turret, together with a co-axial 7.7mm (0.303in) BESA machine-gun; later variants also had a second BESA in the hull front. The tank typically mounted a No.9 wireless radio set.

The Valentine I was powered by a 98kW (131hp) AEC A-189 petrol engine, while subsequent marks had diesel engines. Its running gear were distinctive; there were two sets of bogie-mounted treble road-wheels (two small, one medium) per side, plus a medium-sized front idler and a very large rear driver. The hull sides extended over the top of the track, acting like shallow side skirts. The tank's reasonable power-to-weight ratio, however, was offset by its narrow tracks; consequently the Valentine's tactical mobility was modest. It had a top road speed of just 24km/h (15mph), and a disappointing maximum by-road operational range of 145km (90 miles).

Up-powered

Later variants were up-gunned and up-powered. The Marks VIII-X mounted the more potent '6-pounder' 57mm (2.24in) OQF gun in a welded turret. All Valentines from the Mark IV onward (save for the Mark VI) were powered by a 98kW (131hp) GMC-6004 diesel engine. The Soviets found the Valentine's simple engineering suitable to the Eastern Front's conditions. That said, the '2-pounder' gun was barely adequate and Soviet factories replaced some of these weapons with their own 76.2m (3in) tank cannon.

Valentine Mk VII

Crew 3
Production 1941–42
Weight 17.3 tonnes (19.1 tons)
Dimensions 5.41m (17ft 9in) x 2.63m (8ft 8in) x 2.27m (7ft 6in)
Engine 98kW (131hp) A-190 diesel
Road Speed 24km/h (15mph)
Range 145km (90 miles)
Armament '2-pounder' 40mm (1.57in) OQF L/52 gun; 2 x 7.7mm (0.303in) BESA machine-gun, 1 co-axial, 1 hull front
Armour 8–65mm (0.32–2.56in)

Infantry Tank Mk.IV Churchill

The third model of British-supplied infantry support tanks that the Red Army operated during their struggle against the Axis forces on the Eastern Front was the heavy and cumbersome Churchill.

Churchill II

One of just a few Mark IIs in Red Army service, this unusual tank lacks the standard track cover plates.

During the period 1942–1944, the Western Allies dispatched 346 British-designed Infantry Tank Mark IV (A22) Churchill tanks to the Soviet armed forces, of which 280 arrived successfully. The Red Army employed three Churchill tank variants, the majority of which were Mark III and Mark IV vehicles. The Churchill was a heavily-armoured, slow and relatively immobile heavy infantry support tank, and it had a distinctive appearance. It was a squat and shallow coffin-shaped vehicle, with the hull superstructure contained largely within the space between the tracks; the hull-top decking was thus barely higher in the vehicle than the tops of the tracks. The Mark I and II tanks had a completely horizontal hull deck, while the Churchill Mark III and IV tanks had slightly raised shallow 'wedges' on their hull deck front and rear sections.

Weighing a considerable 39.9 tonnes (43.8 tons), the five-man Churchill II mounted the '2-pounder' 40mm (1.57in) OQF L/52 gun in a rounded-edged cuboid turret that featured a small rectangular stowage box on the rear face. The tanks also featured one co-axial and one hull front-mounted 7.7in (0.303in) BESA machine-guns.

The Churchill was also heavily protected. The Marks I–IV had armoured plates of 16–102mm (0.63–4.02in) thickness. The Soviets received just 26 of the 47 Mark II tanks dispatched during 1942; while they appreciated the Churchill's high battlefield survivability, the lethality of the '2-pounder' was disappointing.

TACTICAL EMPLOYMENT

The Soviets organized the Churchill tanks they received from the British into ten separate Guards Armoured Breakthrough Regiments, numbered in the range 10th–82nd Regiments. Each unit was equipped with 21 Churchills and staffed by 214 personnel. During 1942–44 these regiments fought in operations to lift the Siege of Leningrad, the battle to surround Stalingrad, the titanic clash of armour at Kursk and the liberation of Kiev and Orel.

Churchill II
Crew 5
Production 1941–43
Weight 39.9 tonnes (43.8 tons)
Dimensions 7.42m (24ft 5in) x 3.45m (11ft 4in) x 2.74m (9ft)
Engine 261kW (350hp) Bedford Twin-Six petrol
Road Speed 24km/h (15mph)
Range 140km (88 miles)
Armament '2-pounder' 40mm (1.57in) OQF L/52 gun; 2 x 7.7mm (0.303in) BESA machine-gun, 1 co-axial, 1 hull front
Armour 16–102mm (0.63–4.02in)

Churchill Mk III

The subsequent Churchill Mark III and IV tanks variants – 301 examples of which served with the Red Army from 1942 onwards – were up-gunned models. These variants mounted the more potent '6-pounder' 57mm (2.24in) OQF gun in modified turrets. The turret fitted to the Mark III was

larger and square-edged with a redesigned rear stowage box, whereas the enlarged turret of the Mark IV had rounded edges.

Churchill Mark I–IV tanks were powered by a 261kW (350hp) Bedford Twin-Six petrol engine. The tank's track and suspension arrangement was also distinctive. The vehicle had nine independent very small road-wheels per side attached to vertical coil springs, plus one front and one rear similarly-sized 'spacer' wheels; both the driver and idler wheels were very large. The hull sides on the Churchill extended down over the track tops, acting as shallow side skirts.

The power plant, transmission and running gear enabled this heavy tank only to attain a modest maximum speed of just 24km/h (15mph); moving tactically across country, the tank could not even obtain half this speed. Unsurprisingly, given its 'gas-guzzling' engine, the cumbersome Churchill could also only achieve a modest maximum operational range of 140km (88 miles) when travelling on road.

Churchill IV
Crew 5
Production 1942–43
Weight 39.9 tonnes (43.99 tons)
Dimensions 7.42m (24ft 5in) x 3.45m (11ft 4in) x 2.74m (9ft)
Engine 261kW (350hp) Bedford Twin-Six petrol
Road Speed 24km/h (15mph)
Range 140km (88 miles)
Armament '6-pounder' 57mm (2.24in) OQF; 2 x 7.7mm (0.303in) BESA machine-gun, 2 x 7.7mm (0.303in) BESA machine-gun, 1 co-axial, 1 hull front
Armour 16–102mm (0.63–4.02in)

Churchill III
Both the thicker barrel of the '6-pounder' 57mm (2.24in) OQF gun and the hull decking 'wedges' identify this as a Mark III. This one is from the 48th Separate Heavy Guards Breakthrough Tank Regiment, which was involved in the liberation of Kiev in November 1943.

Churchill IV
This vehicle shows an interesting mix of original white-painted English language instructions and subsequently-added Soviet battle slogans.

Light Tank Mk.VII Tetrarch

Driven overland into the southern Caucasus region of the Soviet Union, a small number of British-manufactured Tetrarch light tanks assisted the Red Army's north-westward counter-offensive mounted from the Caucasus Mountains' northern foothills.

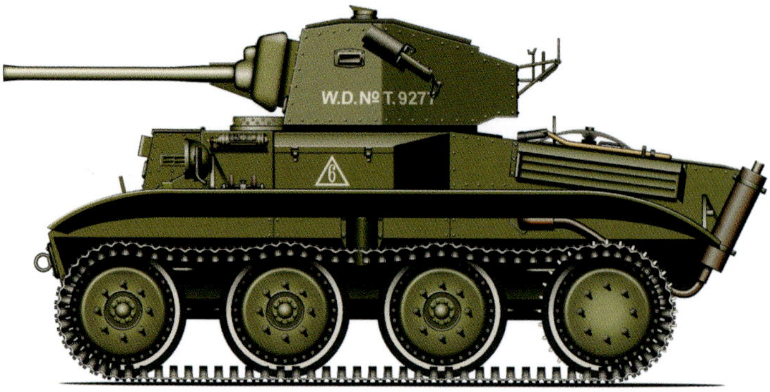

Tetrarch

Perhaps the Tetrarch's most striking feature was its eight large-sized road-wheels with the tracks running tightly around them.

Light Tank Mk.VII Tetrarch
Crew 3
Production 1940–41
Weight 7.6 tonnes (8.4 tons)
Dimensions 4.11m (13ft 6in) x 2.31m (7ft 7in) x 2.12m (6ft 11in)
Engine 123kW (165hp) Meadows 12-cylinder petrol
Road Speed 64km/h (40mph)
Range 230km (143 miles)
Armament '2-pounder' 40mm (1.57in) OQF L/52 gun; 1 x co-axial 7.7mm (0.303in) BESA machine-gun
Armour 4–14mm (0.16–0.55in)

During Winter 1941–42, the British Army dispatched 20 Vickers-designed Light Tank Mark VII (A17) Tetrarch vehicles overland up to north-western Persia (modern Iran), and then on into the southern Caucasus region of the Soviet Union. These vehicles were taken from the strength of the 1st British Armoured Division, then deployed in Egypt. The Soviets found that this light fully-tracked armoured vehicle was well suited for patrolling duties across the open and rugged hilly terrain of the region; given the tank's limited survivability, however, it would have been almost suicidal for their crews to engage enemy armour.

DEPLOYMENT DETAILS

In early January 1942 the 20 Tetrarchs reached Soviet territory. All the tanks came from the 1st Armoured Division's 10th Royal Hussars. During spring 1942 Soviet crews from the 21st Independent Training Tank Regiment, based in Armenia, familiarized themselves with these tanks. Next, the Tetrarchs conducted patrolling duties in the Southern Caucasus. Finally, in early 1943 the tanks participated in the Soviet counter-offensive against the German forces that had advanced into the Northern Caucasus.

Reconnaissance role

The Mk.VII Tetrarch was a small fully-tracked light tank. In terms of external visual appearance it looked very much like a tracked version of an eight-wheeled armoured car, but with a rather oversized turret. In the latter's front face the vehicle mounted a '2-pounder' 40mm (1.57in) OQF L/52 gun, for which it carried 40 rounds of ammunition. The turret also featured a co-axial 1 x 7.7mm (0.303in) BESA machine-gun, for which 2025 rounds were stowed. Weighing a mere 7.6 tonnes (8.4 tons), the three-man Tetrarch had only very light protection, with armoured plates that were just 4–14mm (0.16–0.55in) thick.

The vehicle was powered by a rear-located 123kW (165hp) Meadows 12-cylinder petrol engine, which enabled the Tetrarch to achieve an impressive top road speed of 64km/h (40mph). The Tetrarch had four large road-wheels per side, around which the track ran tightly, with neither driver/idler wheels nor return rollers. On top of this sat a rectangular hull. Even though the vehicle only had small fuel tanks, it could still obtain a decent maximum operational range of 230km (143 miles).

Vickers Mk VII

The Soviet North-Caucasus and Trans-Caucasus Fronts immediately put these vehicles – dubbed the Vickers Mark VII by the Red Army – to use in armoured recce roles and patrolling actions in the rugged foothills of the Caucasus Mountains. Subsequently, many of these 20 Tetrarch tanks participated in the counter-offensive actions mounted during the spring of 1943 by the 151st Tank Brigade – part of the North-Caucasus Front's subordinate Forty-Fifth Army.

Light Tank M3 Stuart

Just under 1000 American-supplied M3 Stuart light tanks served in Red Army armoured units during the middle phase of the latter's struggle to eject the hated Axis invaders from Soviet soil.

M3 Stuart

In Soviet service near Volkhov in May 1942, a second production batch M3 with welded turret and riveted hull.

M3 Stuart

Crew 4
Production 1941–42
Weight 12.9 tonne (14.2 ton)
Dimensions 4.45m (14ft 8in) x 2.22m (7ft 4in) x 2.30m (7ft 7in)
Engine 186kW (250 hp) Continental 7-cylinder radial petrol or Guiberson T-1070 diesel
Road Speed 58km/h (36mph)
Range 120km (75 miles)
Armament 1 x 37mm (1.45in) M5 or M6 L/56.6 gun; 4 or 5 x 7.62mm (0.3in) Browning M1919A4 machine-guns, 1 co-axial, 1 hull front, 2 hull sponsons, sometimes 1 turret roof
Armour 13–51mm (0.52–2in)

Developed during 1939–41, the American M3 General Stuart (often just termed the 'Stuart') light tank was an evolutionary development of its predecessor, the M2A4 light tank. Subsequently, during March 1941 the American Car and Foundry Company began production of the four-man Stuart tank.

Between late 1941 and mid-1942, the American dispatched 1300 M3 Stuarts – a considerable number of them petrol-engined vehicles – by ship to the Soviet Union. Many were lost to German submarine and aircraft attack en route, so only 900 Stuarts reached Soviet soil.

High profile

The 12.9 tonne (14.2 ton) M3 Stuart was a fairly small but relatively tall light tank; given its weight the tank was clearly at the upper end of the classification spectrum. The Stuart was a mere 4.45m (14ft 8in) long and also equally narrow at just 2.22m (7ft 4in). Yet by 2.30m (7ft 7in) it was slightly taller than it was wide – a relatively rare configuration in tank design. Its weight, therefore, came more from the vehicle's level of protection rather than its inherent size.

In terms of physical external appearance, the most striking feature was the sheer height of the lozenge formed by the track; this was some 40 per cent of the height of the entire vehicle. On top of this was a relatively tall box-like rectangular riveted hull superstructure. Positioned forward upon this superstructure was the tank's angular octagonal-shaped large and tall welded turret. The latter was distinguished by a large and tall, angular commander's cupola, which was set in the left rear of the turret roof. The turret front narrowed to create a relatively small rectangular face, which was riveted to the rest of the welded turret structure.

In the front face of this turret, the Stuart mounted a narrow circular gun mantlet. Projecting from this was the tank's main armament, the 37mm

M3A1 STUART

From May 1942 Stuart production transited from the M3 to the significantly modified M3A1. This new variant featured a re-designed rounded-edged turret that incorporated an internal turret basket and lacked the commander's cupola. In the M3A1 the two 7.62mm (0.3in) Browning M1919A4 machine-guns mounted in the sponson fitted to each side of the central hull upper superstructure were also discontinued. Some 4621 vehicles of the final variant – the M3A3 (Stuart Mark V) – were manufactured until February 1943.

(1.46in) M5 or M6 L/56.6 gun. Early Stuart tanks sported no fewer than four or five 7.62mm (0.3in) Browning M1919-A4 machine-guns: one was co-axial to the right of the M6 cannon; one was mounted in the front face of each of the two box-like sponsons fitted to the right and left sides of the central hull superstructure; one was ball-mounted in the right hull front; and sometimes there was another one fitted to an M20 anti-aircraft mount on the very rear of the cupola. For a light tank the Stuart featured relatively high levels of protection: its armoured plates varied in thickness from 13mm (0.52in) to 38mm (1.5in) on the hull front and a considerable 51mm- (2in-) thick on the gun mantlet.

Track configuration

The Stuart's track and suspension arrangement comprised two pairs per side of open-spoked small-sized road-wheels fitted to elongated Vertical Volute Spring Suspension (VVSS) bogie mountings. The tank had an extremely large open-spoked rear driver wheel and a medium-sized front idler, as well as three very small return rollers. The overall impression is one of the amount of open space within the wheel and track system.

The Stuart derived its power either from a 186kW (250hp) Continental W-670 7-cylinder petrol engine or a Guiberson T-1070 diesel engine (in the Stuart Mark II). This combination of running-gear, Vertical Volute Spring Suspension, power-plant and transmission enabled the M3 Stuart to attain an impressive maximum road speed of 58km/h (36mph). That said, the vehicle possessed relatively small fuel tanks, which restricted its operational range when travelling by road to a disappointing 120km (75 miles).

New turret

Mid-production models of the M3 featured a re-designed rounded horseshoe-shaped turret with an ellipsoid riveted drum-like cupola. Then from May 1942, the first significantly modified variant – the M3A1 (or Stuart Mark III) – entered service.

M3A1 Stuart
Crew 4
Production 1941–43
Weight 13.3 tonnes (14.7 tons)
Dimensions 4.45m (14ft 8in) x 2.22m (7ft 4in) x 2.56m (8ft 5in)
Engine 186kW (250 hp) Continental 7-cylinder radial petrol
Road Speed 58km/h (36mph)
Range 120km (75 miles)
Armament 1 x 37 mm (1.45in) M6 L/56.6 gun; 3 x 7.62mm (0.3in) Browning M1919-A4 machine-guns, 1 co-axial, 1 hull front, 1 turret roof
Armour 13–51mm (0.52–2in)

M3A1 Stuart

A rare early M3A1 with the latter's re-designed turret, with no cupola and turret pistol ports that retained the M3's sponson machine-guns. The slogan daubed on the turret reads: 'We broke the back of the Fascist Beast.'

M3A1 Stuart

The Americans dispatched 340 M3A1 Stuarts to the Soviet Union during the latter half of 1942.

M3 General Lee

The tall-silhouetted M3 General Lee, with its hull sponson-mounted howitzer and turret roof-mounted auxiliary turret, may not have been the Soviets' most popular Lend-Lease tank, but it nevertheless provided useful service during 1942.

During 1941–43 the Western Allies dispatched some 1386 M3 General Lee (or 'Grant') medium tanks via the North Atlantic convoy route or the land route through Persia, of which 1200 arrived. The Soviets used many of these during Operation 'Mars', their summer 1942 offensive west of Moscow.

Weighing 27.2 tonnes (29.9 tons), the tall-silhouetted M3 Lee entered mass production in June 1940. The seven-man tank was protected by armoured plates that varied in thickness from 57mm (2.24in) on the mantlet down to 30mm (1.2in).

Two main armaments

Unusually, the stop-gap Lee design mounted two main armament weapons. In its small rounded turret the tank mounted the 37mm (1.46in) M5 or M6 L/56.6 gun. In the tank's right-hand hull front sponson the Lee also mounted a short-barrelled 75mm (2.95in) M2 or M3 gun. The vehicle also featured two 7.62mm (0.3in) Browning machine-guns, one co-axially to the 37mm (1.46in) cannon and one mounted in a small auxiliary turret atop the main turret.

Power plant

The Lee was powered by a 254kW (340hp) Continental R-975 radial petrol engine, and could obtain a maximum road speed of 42km/h (26mph). On a single tank of fuel it could also attain a maximum operating range of 193km (120 miles).

The M3A1–M3A5 variants, produced from mid-1942 onwards, featured many modifications including cast upper hulls, twinned GMC 6046 diesel engines and a crew reduced from seven to six.

M3 General Lee
Crew 7
Production 1940–43
Weight 27.2 tonnes (29.9 tons)
Dimensions 5.64m (18ft 6in) x 2.72m (8ft 11in) x 3.12m (10ft 3in)
Engine 254kW (340hp) Continental R-975 radial petrol
Road Speed 42km/h (26mph)
Range 193km (120 miles)
Armament 37mm (1.46in) M5 or M6 L/56.6 gun (turret); 75mm (2.95in) M2 or M3 gun (hull front); 2 x 7.62mm (0.3in) Browning M1919-A4 machine-guns, 1 co-axial, 1 auxiliary turret
Armour 30–57mm (1.2–2.24in)

M3 Lee (early production)

Seen on the Volkhov in July 1942, this early M3 sports a riveted hull and large hull-side escape hatches. The tank's armament remains the original 75mm M2 short-barrelled howitzer and the 37mm M5 cannon.

M4A2 Sherman

The M4 Sherman tank became the main Western Allied battle tank of the second half of the Second World War; unsurprisingly over 4000 were dispatched to the Soviet Union and employed there to good effect.

In the period from late 1942 through 1943, approximately 2000 75mm-gunned diesel-powered M4A2 Shermans reached the Red Army. Moreover, during 1944–45 a further 2200 M4A2 76(W) tanks were dispatched to the Soviet Union; a few petrol-engined Shermans were also received. Produced in 18 major variants, the five-man Sherman medium tank became by 1943–44 the Western Allies' standard main battle tank, functioning in the tank killer, exploitation and infantry support roles. Allied factories produced a staggering 56,000 Sherman tanks before the war's end in May 1945.

Distinct profile

In terms of external appearance from the side perspective the tall-silhouetted Sherman featured a tall track lozenge, a wedged-shape hull and a front-sloped cylindrical turret. The track and Vertical Volute Spring Suspension (VVSS) arrangement consisted of three paired bogie-mounted medium-sized road-wheels; the tank also had medium-sized front driver and rear idler wheels, together with three small upper-track return rollers. The wedge-shaped hull superstructure comprised a long sloped hull frontal plate, a short flat hull decking section and a long gently-sloped rear section that tapered down to a shallow slightly sloped rear hull plate.

The Sherman tank was developed during 1940–41. The first five 75mm-gunned production designs – the welded-hull M4, the rounded-edged cast-hull M4A1, the diesel-powered M2A2, the 'wet'-stowage Ford-engined M4A3 and the Chrysler-engined M4A4 tanks – all entered production during a six-month period February–July 1942. Between early 1942 and May 1944, Allied factories produced 9189 examples of these five basic 75mm-gunned Sherman tank types.

M4A2 (early production)

Two early-production 75mm-gunned M4A2 tanks on the Eastern Front; they are most easily identified by the relatively shallow hull frontal plate.

M4A2 (early production series)
Crew 5
Production 1942–43
Weight 28.4 tonnes (31.3 tons)
Dimensions 5.92m (19ft 5in) x 2.62m (8ft 7in) x 2.74m (9ft)
Engine 306kW (410hp) General Motors GM-6046 'twin' 2 x 6-cylinder diesel
Road Speed 39km/h (24mph)
Range 193km (120 miles)
Armament 75mm (2.95in) M3 L/40; 2 x 7.62mm (0.3in) Browning M1919-A4 machine-guns, 1 co-axial, 1 hull front
Armour 38–76mm (1.5–2.99in)

These Shermans mounted the relatively short-barrelled 75mm (2.95in) M3 L/40 gun in a half-cylindrical mantlet fitted to the sloped-front of the vehicle's fully-revolving turret. The tank stowed 90 rounds for this gun. Firing the M61 Armour-Piercing Capped (APC) round at a muzzle velocity of 617m/s (2024ft/s) the 75mm-gunned Sherman could penetrate 86mm (3.3in) of vertical armour at a range of 1000m (3280ft). The Sherman also mounted one 7.62mm (0.3in) Browning machine-gun co-axially with the main gun, and a second Browning in a ball-mount in the right hull front plate.

The Sherman featured reasonable levels of protection by the standards of 1942–43. The turret's steel frontal armour was up to 76mm (2.99in) thick, as was that on the gun mantlet. The turret sides and rear consisted of 50–64mm- (1.97–2.52in-) thick armour. The tank's upper hull frontal plate was 51mm (2in) thick and angled at 56 degrees in early-production variants and at 47 degrees in later versions. Finally, the hull sides were protected by 38–45mm- (1.5–1.77in-) thick vertical armour. By 1944, however, as AFV-lethality continued to increase, this level of protection had become relatively unimpressive, and the tendency of Shermans to burn fiercely when hit became a significant concern.

Weighing 28.4 tonnes (31.3 tons), early Shermans were powered by a 298kW (400hp) Continental R975-C1 nine-cylinder or a General Motors GM-6064 diesel engine. The combination of this power-plant, VVSS suspension and the Spicer synchromesh transmission enabled the Sherman to obtain a maximum road speed of 39km/h (24mph). Fitted with 560 litre (148 gal) internal fuel tanks, with a single load of fuel it could achieve a modest maximum by-road operational range of 193km (120 miles).

Lend-Lease vehicles

Some 2000 M4A2 Shermans served with the Red Army on the Eastern Front. Of these about half were early-production vehicles and half were late-production types. The early-production M4A2 tanks were distinguished from the other variants by their slightly more shallow sloped – at 56 degree from the vertical – hull frontal plate. The late-production M4A2's in Soviet service featured a more steeply sloped (47 degree) frontal plate and hull armour thickened in places.

From early 1944 onwards until May 1945 a proportion of Sherman production was shifted to manufacture tanks that mounted the longer-barrelled 76.2mm (3in) M1A1 L/52 gun. There were three main types of 76mm-gunned Shermans, the M4A1 76(W), M4A2 76(W) and M4A3 76(W). These tanks featured a re-designed T23 turret. They also carried reduced stowage of 71 main gun rounds, housed in 'wet' conditions to minimize fires.

VVSS vs. HVSS

All early Sherman tanks featured the Vertical Volute Spring Suspension (VVSS) arrangement for their three bogie-mounted paired medium-sized road-wheels on each side. Here the suspension spring sat vertically behind the bogie assembly. However, a proportion of later Sherman production tanks instead featured the alternate Horizontal Volute Spring Suspension (HVSS). In this arrangement, the spring was fitted horizontally between the two suspension arms of the bogie-mounting; this spring, therefore, was located just above the paired road-wheels.

M4A2 (late production series)
Crew 5
Production 1943–44
Weight 31.8 tonnes (35.1 tons)
Dimensions 5.92m (19ft 5in) x 2.62m (8ft 7in) x 2.74m (9ft)
Engine 306kW (410hp) General Motors GM-6046 'twin' 2x 6-cylinder diesel
Road Speed 39km/h (24mph)
Range 240km (150 miles)
Armament 75mm (2.95in) M3 L/40; 2 x 7.62mm (0.3in) 2 x Browning M1919-A4 machine-guns, 1 co-axial, 1 hull front
Armour 13–108mm (0.5–4.25in)

M4A2 (late production)
A late-production M4A2 sporting the short-barrelled 75mm (2.95in) gun, T54E1 steel tracks and the steeper hull frontal plate.

M4A2 76(W)

Operating near Berlin in May 1945, this 76mm-gunned Sherman is identified by the redesigned T23 turret and the longer-barrelled gun.

Some 2073 M4A2 76(W) tanks operated in Soviet service during late 1944 and early 1945. The 76.2mm (3in) gun was a more potent tank killer than the 75mm (2.95in) Sherman gun, but had the detriment of reduced high-explosive (HE) capability and increased muzzle blast that complicated target acquisition. A proportion of Sherman 76(W) manufactured also produced vehicles that utilized the alternate Horizontal Volute Suspension System (HVSS).

While not always universally popular with their crews, the 4000-plus Sherman tanks that served in the Red Army until the war's end provided a significant boost to the latter's combat power during their Herculean efforts to defeat the despised Axis armed forces.

M4A2 76(W)

This tank entered Brno in Czechoslovakia with the IX Guards Mechanized Corps, 2nd Ukrainian Front, in late April 1945.

M4A2 Sherman 76(W)
Crew 5
Production 1944–45
Weight 33.3 tonnes (36.7 tons)
Dimensions 6.32m (20ft 8in) x 2.67m (8ft 9in) x 2.97m (9ft 9in)
Engine 306kW (410hp) General Motors GM-6046 'twin' 2x 6-cylinder diesel
Road Speed 39km/h (24mph)
Range 161km (100 miles)
Armament 76.2mm (3in) M1A1, M1A1C or M1A2 L/52 gun; 7.62mm (0.3in) Browning M1919-A4 machine-guns, 1 co-axial, 1 hull front; 1 x 12.7mm (0.5in) Browning M2HB heavy machine-gun (turret roof)
Armour 38–76mm (1.5–2.99in)

M4A2 76(W)

Vehicle 936, a 76mm-gunned Sherman fitted with muzzle-brake, in action near Vienna in April 1945.

M10 Wolverine

In one of the later batches of Lend-Lease vehicles to be sent to the Soviet Union, a small number of M10 Wolverine tank destroyers entered Red Army service after travelling along the over-land route through Persia.

M10 Wolverine

This mid-production M10 in Soviet service sports the triangular counter-weight mounted on the turret's upper rear face.

During 1944, the United States dispatched 52 Three-inch Gun Motor Carriage (GMC) M10 Wolverine tank destroyers to the Soviets. Some of these fought with the 1223rd Self-propelled Artillery Regiment, a subordinate unit of the 5th Guards Tank Army. The 29.13 tonne (32.1 ton) M10 Wolverine was a turreted tank destroyer based upon the M4A2 Sherman tank chassis. Developed during 1942–43, the Wolverine entered mass production in September 1943. During late 1943 American factories produced 4993 M10s; the 1700 late-production vehicles completed during 1944 utilized the M4A3 Sherman chassis and were designated the M10A1.

M10 Wolverine
Crew 5
Production 1943
Weight 29.13 tonnes (32.1 tons)
Dimensions 6.83m (22ft 5in) x 3.05m (10ft) x 2.9m (9ft 6in)
Engine 306kW (410hp) General Motors GM-6046 6-71 'twin' 2x 6-cylinder diesel
Road Speed 40km/h (25mph)
Range 322km (200 miles)
Armament 76.2mm (3in) M7 L/50 gun; 12.7mm (0.5in) Browning M2HB heavy machine-gun (turret rear)
Armour 9.5–57.2mm (0.8–2.25in)

Open-roofed turret

The five-man Wolverine mounted an angular five-sided open-roofed turret; in mid- and late-production M10 vehicles (as well as M10A1) the turret rear projected backwards to form a heavy triangular counter-weight to balance the gun's weight. The lack of a roof facilitated situational awareness and maintenance of the gun, but this was bought at the price of the increased vulnerability of the crew.

The M10 mounted the 76.2mm (3in) M7 L/50 gun in a triangular-shaped gun shield fitted to the turret's flat-front face. Late production M10 vehicles, however, instead mounted the 76.2mm (3in) M1 gun. By way of secondary armament, the Wolverine mounted a 12.7mm (0.5in) Browning M2HB heavy machine-gun on the turret rear.

The hull superstructure of the Wolverine had a box-like appearance with a shallow lower section and a taller upper portion. The superstructure's lower section comprised outwardly-sloping plates. Conversely, the hull's upper section comprised inwardly-sloping side and rear armoured plates. The vehicle's hull front comprised a long well-sloped glacis plate. The Wolverine had only modest protection, with its armoured plates varying in thickness from just 9.5mm (0.8in) to 57.2mm (2.25in).

The M10 Wolverine employed the standard Sherman chassis with three bogie mountings that sported a pair of medium-sized road-wheels married to the Vertical Volute Spring Suspension (VVSS) arrangement. Powered by the 306kW (410hp) General Motors GM-6046 6-71 'twin' 2 x 6-cylinder diesel engine (or the 336kW/450hp Ford GAA diesel in the M10A1), the Wolverine could obtain a top road speed of 40km/h (25mph), and a maximum operational road-based range of 322km (200 miles).

Red Army Tanks and Variants

Below is a list by type of all significant tank and armoured fighting vehicles and variants deployed by the Red Army during World War II.

LIGHT TANKS

T-17	One-man, 2.2 tonne (2.4 ton), light reconnaissance tankette that mounted a 7.62mm (0.3in) DT machine-gun.
T-18 (MS-1)	First ever Soviet-produced tank, entering service in 1925, a modification of the French Renault FT light tanks captured during the civil war; the T-18 mounted a short-barrelled 37mm (1.46in) Model 1928 gun, and featured 8–16mm (0.31–0.63in) thick armoured plates.
T-18M	Desperate late-1941 expedients, these modified old T-18s had their unreliable engines removed to make space for the larger rounds required for the newly-mounted 45mm gun (1.77in) 20K gun.
NI 'Odessa Tank'	Desperate mid-1941 expedient, the Odessa factory placed rough armoured superstructures onto 42 existing military tractors and added either an available T-26, T-37A or T-38 turret.
T-19	Unsuccessful three-man light tank design that mounted the 37mm (1.46in) Model 1930 gun and which featured sloped armour; intended to replace the T-18, production of this variant was cancelled after just 14 vehicles had been built.
T-26 Model 1931	A double-turreted, petrol-engined, 8.2 tonne (9.13 ton) light tank manufactured during 1931–33 that had a distinctive 'wedged-shaped' track arrangement, and which featured thin 6–15mm (0.24–0.59in) armoured plates; each of the tank's side-by-side arranged cylindrical turrets mounted a 7.62mm (0.3in) DT machine-gun.
T-26 Model 1931 (up-gunned)	Rare variant of the Model 1931, this 8.7 tonne (9.6 ton) tank mounted a 37mm (1.46in) PS1 gun in the right-hand turret instead of the standard 7.62mm (0.3in) DT machine-gun.
T-26 Model 1932	This variant incorporated minor improvements to the standard machine-gun equipped double-turreted Model 1931 T-26 design; it also retained the 66kW kW (88hp) GAZ-T26 4-cylinder petrol that had powered the Model 1931.
T-26TU Model 1931	Command tank variant of the double-turreted T-26 Model 1931 fitted with a wireless radio set and a hull top-mounted horizontal rail antenna; the vehicle mounted a 37mm (1.46in) PS1 gun in its right-hand turret.
T-26 Model 1933	Single-turreted main production petrol-engined T-26 variant, weighing 9.4 tonnes (10.36 tons), which featured the more potent longer-barrelled 45mm (1.77in) 20K Model 1932 gun; nearly 3500 examples of this variant were produced during 1933–35.
T-26RT Model 1933	Command tank variant of the single-turreted T-26 Model 1933 that featured a wireless radio set and a horseshoe-shaped turret antenna.
T-26 Model 1936	The main production variant during 1936–38, this 9.65 tonne (10.6 ton) tank incorporated many minor design developments to speed manufacture; it also mounted the new Model 1934 45mm (1.77in) gun as well as three 7.62mm (0.3in) DT machine-guns.
T-26 Model 1938	This 9.8 tonne (10.8 ton) variant featured a redesigned conical turret with sloped sides that featured the new, Model 1938 version, of the 45mm (1.77in) 20K gun; it was powered by an up-rated 71kW (95hp) GAZ-T26 4-cylinder petrol engine.

T-26 Model 1939 This late-production up-armoured 10.25 tonne (11.3 ton) T-26 variant, manufactured during 1939–41, featured a better-protected re-designed conical turret and a modified hull and superstructure; it featured 6–20mm (0.24–0.79in) armoured plates.

T-26E Model 1939 ('Ekranami') This 12 tonne (13.3 ton) T-26 variant was up-armoured during 1940 through the fitting (or retro-fitting on existing T-26s) of 30–40mm (1.18–1.57in) bolted-on appliqué armoured plates on top of the design's standard 6–20mm (0.24–0.79in) thick armour.

KhT-26 (OT-26) Petrol-engined flame-throwing ('chemical tank') variant of the double-turreted T-26 Model 1931; with a long-barrelled flame-device mounted in the right-hand turret, and with the left turret removed.

KhT-130 (OT-130) Flame-throwing variant of the single-turreted T-26 Model 1933 tank, with a short-barrelled flame-device replacing the 45mm (1.77in) gun.

KhT-133 (OT-133) Flame-throwing tank variant of the T-26 Model 1938, which featured a short-barrelled flame-device instead of the standard 45mm (1.77in) gun.

T-26-T2 Modified T-26 tank chassis intended for towing field artillery pieces.

TTs-26 A fuel-carrying variant that carried a large cylindrical fuel tank on the T-26 chassis.

AT-42 Fully-tracked heavy artillery tractor based on the T-26 tank chassis.

T-27 A 2.5 tonne (2.7 ton) lightly-protected tankette that mounted a 7.62mm (0.3in) DT machine-gun, which was produced during 1931–34.

T-37A Amphibious light tank powered by a 30kW (40hp) GAZ-AA petrol engine and sporting 3–10mm (0.12–0.39in) thick armoured plates; atop the hull centre-right the tank featured a tall cylindrical turret that mounted a 7.62mm (0.3in) DT machine-gun.

T-37TU This 3.21 tonne (3.53 ton) Platoon Commander's variant of the T-37A, in production during 1934–36, sported a wireless radio set and a horizontal rail antenna mounted along its hull top.

T-38 Built during 1936–39, this thinly-armoured, flat-hulled 3.3 tonne (3.6 ton) air-portable amphibious light scout tank mounted a 7.62mm (0.3in) DT machine-gun as its main armament; it had two pairs of bogie-mounted open-spoked road wheels per side.

T-38RT An up-gunned variant produced in small numbers during 1938–39; instead of a DT machine-gun the vehicle mounted the quick-firing long-barrelled 20mm (0.79in) ShVAK cannon.

T-40 A squat, rectangular, tall-hulled 5.5 tonne (6.1 ton) amphibious light scout tank produced during 1939–41, which mounted a 12.7mm (0.5in) DShK heavy machine gun and a co-axial 7.62mm (0.3in) DT machine-gun; it also featured thin 3–13mm (0.12–0.51in), but slightly-sloped, armoured plates.

T-40S A rare non-amphibious 'land-only' variant of the T-40 produced during 1940–41 that featured thicker turret armour.

BM-8-24 [T-40] During 1941 the Soviets converted 44 T-40 tanks into the BM-8-24 rocket rocket-launcher vehicle; the BM-8 launcher had 24 rails for its 82mm (3.23in) rockets.

T-46 A series of four slightly different mid-1930s prototypes that incorporated the Christie suspension into a modified T-26 chassis.

T-50 This promising but complexly-engineered 13.7 tonne (15.1 ton) diesel-powered light tank proved too costly to manufacture and so production was halted after just 69 vehicles had been completed; it mounted a 45mm (1.77in) Model 1938 gun.

T-60 (early production)	Manufactured during 1941–42, this 5.2 tonne (6.1 ton) two-man petrol-engined light tank featured well-sloped, albeit still light, armoured plates; its hexagonal angular turret mounted a 20mm (0.79in) TNSh L/82.4 automatic-fire cannon and a co-axial 7.62mm (0.3in) DT machine-gun.
T-60 (late production)	Later T-60 tanks had their frontal armoured plates increased to 35mm (1.38in) thickness, and featured solid-disc (rather than open-spoked) road wheels.
BM-8-24 [T-60]	During late 1941 the Soviets removed the turrets of a small number of T-60 tanks and instead mounted the 24-rail BM-8 rocket launcher system.
T-70	Two-man design based on the T-60, this low-silhouetted light tank sported a potent 45mm (1.77in) L/46 Model 1938 gun and well-sloped armoured plates; it was unusual in that it was powered by the unsuccessful expedient of two 'mated in series' 52kW (70hp) GAZ-202 6-cylinder petrol engines.
T-70M	This improved T-70 variant incorporated two engines that were 'mated in parallel', enhanced running gear components, and a modified turret with flat well-sloped armoured plates.
T-80 Model 1943	A development of the T-70, this 11.6 tonne (12.8 ton) variant featured an enlarged two-man turret that mounted the 40mm (1.57in) L/46 Model 1938 gun; production was halted after just 120 vehicles in late 1943.
FAST AND MEDIUM TANKS	
T-24 Medium	First Soviet-designed medium tank, this mechanically-unreliable, petrol-engined, multi-turreted 18.2 tonne (20.1 ton) five-man vehicle mounted a 45mm (1.77in) Model 1932 gun.
T-28 Medium Model 1934	Long, multi-turreted, petrol-engined medium tank; this 28 tonne (30.86 ton) vehicle featured one main turret that mounted a 45mm (0.77in) 20K Model 1932 gun, as well as two auxiliary ones that each had a 7.62mm (0.3in) DT machine-gun.
T-28 Medium Model 1938	Modified late-production variant that mounted the longer-barrelled 76.2mm (3in) L-10 gun; many of these variant also incorporated a turret top mounted 7.62mm (0.3in) DT anti-aircraft machine gun.
T-28E ('Ekranami') (or T-28M)	Up-armoured late-production T-28 variant; existing vehicles returned to the factory during 1939–40 received additional bolt-on appliqué armoured plates.
BT-1	Experimental fast tank prototype based on the novel Christie coil-spring suspension that lacked a turret.
BT-2	First Soviet-designed fast tank, this three-man, 10.2 tonne (11.24 ton), vehicle's use of the Christie suspension and a 298kW (400hp) M-5 petrol engine enabled it to obtain a top speed of 100km/h (62mph); it mounted a 37mm (1.46in) Model 1930 gun and possessed 6–13mm (0.24–0.51in) armour.
BT-2 [machine-gun only]	A small proportion of BT-2 tanks mounted double-barrelled 7.62mm (0.3in) DP machine-guns instead of the usual 37mm (1.46in) gun (due to shortages in the latter); but these vehicles also retained the standard co-axial 7.62mm (0.3in) DT machine-gun.
BT-5 Model 1933	New three-man fast tank design, which married the BT-2 hull to a re-designed turret that mounted a more potent 45mm (1.77in) Model 1932 gun.
BT-5TU Model 1934	Command tank variant of the BT-5 fast tank with 71-TK-1 wireless set and horseshoe-shaped antenna.

BT-5A	An artillery tank variant produced in small numbers that mounted a 76.2mm (3in) howitzer.
PT-1A	An amphibious variant of the BT-5 manufactured in small numbers.
BT-7 Model 1935	Pre-production series of new three-man fast tank design that married the existing BT-5 turret and its 45mm (1.77in) gun to a modified chassis; it featured 6–15mm (0.24–0.59in) armoured plates.
BT-7 Model 1937	Main BT-7 production series that incorporated a new conical turret; this 14 tonne (15.4 ton) vehicle was powered by the up-rated 373kW (500hp) Milukin M17-T V-12 petrol engine.
BT-7 Model 1939	Late-production variant with minor modifications included improved gun sighting systems.
BT-7TU Model 1937	Command tank variant of the BT-7.
BT-7A	Turreted self-propelled howitzer ('artillery tank') version of the BT-7 that mounted a 76.2mm (3in) howitzer (produced 1936–38).
BT-7ATU	Rare command tank variant of the BT-7A.
TT-BT-7	Remote controlled vehicle version of the BT-7.
BT-8	Prototype diesel-engined variant of BT-7 with modified hull that mounted a 45mm (1.77in) 20K Model 1938 gun.
BT-7M	Main production series of the three-man BT-8 that was manufactured during 1939–41; this 14.5 tonne (16 ton) vehicle was powered by the 336kW (450hp) liquid-cooled V-12 diesel engine.

T-34 MEDIUM TANKS

A-20	Novel 20.2 tonne (22.3 ton) prototype design with well-sloped armour that utilized the radical Christie suspension and was powered by a 373kW (500hp) Model V-2 diesel engine; this design led to the development of the T-34.
A-32	Up-gunned and up-armoured development of the A-20, this prototype mounted a 76.2mm (3in) gun, the calibre used in the T-34; it abandoned the little-used and weight-consuming capability to move on road without tracks.
A-34	A development of the A-20 and A-32 designs, this prototype ultimately entered mass production as the T-34 tank.
T-34 Model 1940	First T-34 main production variant; this 26 tonne (28.7 ton) four-man vehicle mounted the 76.2mm (3in) L-11 L/30.5, featured well-sloped 15–45mm (0.59–1.77in) armoured plates, and was powered by the 373kW (500hp) Model V-2-34 diesel engine.
T-34 Model 1940 [Command Tank]	Around one-tenth of Model 1940 production delivered the tank command variant (which shared the same designation as the standard tank); these mounted a 9-RS wireless radio set and a whip antenna.
T-34 Model 1941	Up-armoured and up-gunned T-34 variant; this 26.5 tonne (29.2 ton) four-man vehicle mounted the longer-barrelled 76.2mm (3in) F-34 L/40.5 gun, and featured well-sloped 15–60mm (0.59–2.4in) thick armour.
T-34M	Developmental design work to investigate the feasibility to producing an improved up-gunned T-34/76 that was abandoned in late 1941; influenced the subsequent development of the T-44.

T-34 Model 1942	This 28.5 tonne (31.4 ton) variant incorporated many minor modifications to the basic Model 1941 design, including its maximum armour thickness being increased to 70mm (2.76in); it also carried rectangular box-like external fuel tanks.
T-34 Model 1942 (UZTM)	Many UZTM-built Model 1942 T-34 tanks featured a distinctive rounded-edged 'stamped' pressed two-man turret made from 45mm (1.77in) hardened steel, rather than the 'standard' hexagonal sharp- or soft-edged cast or welded turret of the Model 1941.
T-34 Model 1942 (STZ)	STZ-produced Model 1942s featured the 'traditional' sharp-edged or soft-edged hexagonal cast or welded two-man turret.
T-34 Model 1941/42	STZ-produced T-34 that incorporated some features of both the Models 1941 and 1942.
T-34 Model 1943 (early production)	This 30.9 tonne (34.1 ton) variant was very similar to the Model 1942, with the same gun, armour, engine, and performance figures, but incorporated minor design modifications including twin 'Mickey Mouse' flush-level turret hatches and cylindrical external fuel tanks.
T-34 Model 1943 (mid-production)	Variant where a drum-shaped commander's cupola on the turret roof replaced the flush-level hatch.
T-34 Model 1943 (UTZM)	UTZM-produced Model 1943 with 'stamped' pressed rounded-edged turret.
T-34 Model 1944	Final 76.2mm-gunned T-34 production batches with radio set mounted inside the turret rather than the hull.
T-34/57 Model 1940/ 1941 'Exterminator'	Around 42 of these tank destroyer variants were produced; they mounted the potent 57mm (2.24in) ZiS-4 L/69.3 anti-tank gun.
T-34/57 Model 1943 'Exterminator'	Rare tank destroyer version that mounted the modified 57mm (2.24in) ZiS-4M L/69.3 anti-tank gun.
PT-34 Mine Roller Tank	Engineer tank variant of T-34 Model 1942 fitted with a rotating mine-clearing 'trawl' device, mounted between two jib-booms, which was attached to the vehicle's front hull; it retained its 76.2mm (3in) gun.
PT-34 Mine Roller Tank [T-34/85]	Engineer tank variant of the T-34/85 tank, fitted with a rotating mine-clearing 'trawl' device, which also retained its 85mm (3.35in) main gun.
T-34/85 Model 1943	Interim new diesel-engined T-34 model that mounted the more potent 85mm (3.35in) D-5T L/51.6 gun in a new three-man curved-sided 'egg-shaped' rounded-edged turret, produced from late 1943.
T-34/85 Model 1944 (Standard Turret)	Main production run that mounted the 85mm (3.35in) ZiS-53 or ZiS-S-53 L/51.5 gun.
T-34/85 Model 1944 (Flattened Turret)	Factory 183 produced T-34/85 tank with the slight bulge in the standard turret side's bottom rear having been flattened.
T-34/85 Model 1945	Minor design modifications incorporated into the 1944 design including an enlarged commander's cupola.
OT-34 Model 1942	Flame-throwing variant converted during 1942 from T-34 Model 1942 tanks; mounted the new ATO-41 flame weapon in the right hull front in place of the standard bow 7.62mm (0.3in) DT machine-gun; it retained the 76mm (3in) gun as its main armament.
OT-34 Model 1943	T-34 Model 1943 tanks converted into a flame tank variant that featured an improved ATO-42 flame device in lieu of the bow machine-gun.
OT-34/85	Flame-throwing variant converted from T-34/85 Model 1944 tanks; mounted the new ATO-42 flame weapon in the right hull front in place of the standard bow 7.62mm (0.3in) DT machine-gun; retained the 85mm (3.35in) main gun.

T-34T	Armoured recovery vehicle based on the T-34 chassis that mounted a crane and/or winch.
AT-45	Fully-tracked heavy artillery tractor based on the T-34 chassis.
T-44	This late-war 31.9 tonne (35.2 ton) up-armoured medium tank featured a redesigned T-34/85 turret (that mounted the 85mm [3.35in] D-5T gun) fitted on top of a redesigned shallower chassis that had a novel transverse-mounted engine; it abandoned the Christie suspension in favour of torsion bars.
T-44/100	Unsatisfactory prototype that mounted a more powerful 100mm (3.93in) gun in the T-44 design; this developmental work influenced the creation of the post-war T-54.
HEAVY TANKS	
T-35 Pre-production Series	Pre-production series of the Soviet first indigenous heavy tank design, a five-turreted, 11-man heavy yet relatively lightly-armoured tank; the main turret mounted a 76.2mm (3in) Model 1927/32 gun, while the four auxiliary turrets each mounted either a 45mm (1.77in) 20K Model 1932 gun or a 7.62mm (0.3in) DT machine-gun.
T-35 Model 1936	Main production series of just 35 vehicles, weighing some 43 tonnes (47.4 tons), which were very similar to pre-production series; experience showed that the T-35 was tactically problematic and difficult to manoeuvre.
T-35 Model 1939	Final production series of a slightly modified T-35 variant manufactured during 1938–39; these tanks featured sloped turret side armoured plates rather than vertical ones.
T-100	Series of seven-man, 58 tonne (63.9 ton), heavy tank prototypes produced during 1938–40 that mounted a 76.2mm (3in) L-11 gun and a 45mm (1.77in) cannon.
SMK	Developed during 1938–40, a prototype seven-man 55-tonne (60.6 ton) twin-turreted heavy tank series that mounted both a 76.2mm (3in) L-11 gun and a 45mm (1.77in) gun.
KV-1 Model 1939	First series of KV-1 heavy tank production; this 43.6 tonne (48.1 ton) five-man tank was protected by 25–90mm (0.98–3.54in) thick armour and featured a large, tall, sloped-sided welded turret that mounted the 76.2mm (3in) L-11 L/32 gun.
KV-1 Model 1940 (early production)	Incorporated minor modifications to the Model 1939 design, including the removal of the two lubricant canisters per side previously fitted on the rear fenders; retained its predecessor design's 473kW (600hp) V-2 12-cylinder diesel.
KV-1 Model 1940 (late production)	Up-gunned version that mounted the more lethal 76.2mm (3in) F-32 gun.
KV-1E (Models 1939, 1940)	Up-armouring factory conversion carried out during 1941–42 on 150 existing KV-1 Model 1939, 1940 and 1941 tanks; bolted soft steel 20mm (0.79in) or 30mm (1.18in) appliqué armoured plates were fitted to the turret front and sides, as well as the hull front.
KV-1 Model 1941 (early production)	Variant that incrementally incorporated a series of design modifications, including new radial-style solid-disc road wheels.
KV-1 Model 1941 (mid/late production)	Up-gunned variant of the early production Model 1941 that mounted the more potent 76.2mm (3in) F-34 or ZiS-5 L/40.5 gun; these changes raised the vehicle's weight to 45 tonnes (49.6 tons).
KV-1 Model 1942 (early/mid-production)	An up-armoured variant of the Model 1941, which featured a reinforced cast turret thickened to 120mm (4.72in) in certain areas as well as increased protection on the hull superstructure; these changes raised the vehicle's weight to 48 tonnes (52.9 tons), which further strained its already over-stressed engine, transmission and suspension.

KV-1 Model 1942 (late production)	Final production variant of the Model 1942 that featured a redesigned hull; the latter's flat (rather than rounded) rear plate sloped outwards at a 50-degree angle.
KV-1S Model 1942	To boost the KV-1's tactical mobility, speed and range, this variant featured weight-saving innovations, including reduced levels of protection, and utilized an up-rated 485kW (650hp) V-2K diesel engine; the vehicle's weight was reduced to 43.4 tonnes (47.8 tons).
KV-1S Model 1943	Final KV-1S production variant that incorporated both redesigned road wheels (that returned to the no-holed solid-disc design of early KV-1s) and horizontal turret- and hull-side hand rails onto which infantrymen being transported could hold.
KV-8	Flame tank variant of the KV-1, it mounted an ATO-41 flame device in the turret, which also mounted a 45mm (1.77in) gun instead of the standard KV-1 76mm (3in) gun.
KV-85	Interim development of KV-1-S manufactured during Autumn 1943; this 46 tonne (50.7 ton) variant sported an enlarged turret that mounted the potent 85mm (3.35in) D-5T L/51.6 gun on top of a modified KV-1-S hull; the latter incorporated frontal armour increased from 82mm (3.2in) to 110mm (4.33in) thickness.
KV-13	A series of 32 tonne (35.3 ton) medium tank prototypes developed during 1942-43 that mounted a 76.2mm (3in) ZiS-5 gun and featured well-sloped armoured plates up to a maximum thickness of 120mm (4.72in).
IS-1	First of the newly-designed Iosef Stalin (IS) heavy tank series; this four-man, 43.4 tonne (47.9 ton), vehicle was powered by an 473kW (600hp) V-2-IS engine, mounted a 85mm (3.35in) D-5T gun in a new enlarged cast turret, and featured well-sloped 30-120mm (1.18-4.72in) thick armour.
IS-2 Model 1943 (early production)	An up-gunned development of the IS-1, this 46 tonne (50.7 ton) heavy tank design mounted the more potent 122mm (4.8in) A-19 gun as well as well-sloped and rounded armour.
IS-2 Model 1943 (late production)	This variant mounted the quicker-firing and slightly more lethal 122mm (4.8in) D-25T gun, as well as a 12.7mm (0.5in) DShK heavy machine gun fitted to the top of the commander's cupola on the turret-roof.
IS-2 Model 1944 (early production)	This up-armoured IS-2 variant had its frontal armour tempered to a greater hardness to augment its survivability.
IS-2 Model 1944 (mid/late production)	This variant incorporated minor hull modifications, including a re-configured hull bow plate and a straightened welded or cast hull.
IS-3	The final vehicle in the IS series, the 45.77 tonne (50.45 ton) IS-3 incorporated a novel shallow and wide well-rounded shot-deflecting hemispherical turret that mounted a 122mm (4.8in) D-25T gun; this turret sat on top of a distinctive new well-sloped hull superstructure that featured the infamous 'pike nose' sloped hull frontal plates.

SELF-PROPELLED GUNS

ZiS-30	An improvised 1941 light tank destroyer that married an open-mounted 57mm (2.24in) ZiS-29 anti-tank gun to the existing Komsomolyets artillery tractor.
KhTZ	An expedient 1941 self-propelled anti-tank gun that mounted a 45mm (1.77in) gun in a simple fixed armoured hull superstructure fitted on top of a KhTZ-3 artillery tractor.
ZSU-37	Self-propelled anti-aircraft gun, based on the chassis and hull on the SU-76 assault gun, which mounted the 37mm (1.46in) 61-K automatic-fire anti-aircraft cannon Model 1939; only 12 had entered service by the war's end.

KV-2 Model 1939	Five-man heavy artillery tank (self-propelled howitzer) with a shallow-hull superstructure atop the KV-1 tank's chassis; its large and tall frontally-sloped heavily-armoured cuboid MT-1 turret mounted a 152mm (5.98in) Model 1938/40 L/20 heavy howitzer.
KV-2 Model 1940	The final KV-2 heavy artillery tank variant, this six-man vehicle featured a re-designed large vertically-fronted box-like MT-2 turret.
SU-76	Mechanically unreliable front-engined three-man light assault gun based upon a modified T-70 light tank chassis; this 10.3 tonne (11.4 ton) vehicle mounted the 76.2mm (3in) ZiS-3Sh L/42.6 howitzer in a lightly-armoured closed-topped rear-located fighting compartment.
SU-76M	Reconfigured SU-76 with mated twin engines run in series (one behind the other) on the left rather than forward-located 'in parallel'; it featured an open-topped fighting compartment.
SU-76i	Improvised four-man assault gun based upon the chassis of captured German Panzer III tanks; mounted the 76.2mm (3in) S-1 gun in a fixed superstructure.
SU-76i [on StuG III]	Improvised assault gun that fitted a basic armoured superstructure upon the chassis of eight captured StuG III assault guns; it mounted the Soviet 76.2mm (3in) S-1 gun.
SU-85	Interim four-man diesel-powered 29.6 tonne (32.6 ton) tank destroyer based upon existing T-34 and SU-122 elements; mounted the 85mm (3.35in) D-5S gun in a shallow sloped-sided forward-located armoured superstructure; it featured 20-45mm (0.78–1.77in) armoured plates.
SU-85M	Late-production variant from late 1944 based upon the SU-100's up-armoured chassis; incorporated a modified commander's cupola and a re-designed gun mantlet.
SU-100	More heavily armoured development of the SU-85; this four-crew 31.6 tonne (34.8 ton) vehicle mounted the 100mm (3.94in) D-10S L/53.5 gun in a forward-located well-sloped armoured superstructure and was protected by 20–75mm (0.78in–2.95in) armoured plates.
SU-100Y	A 1940 prototype with no connection to the later SU-100 vehicle; this 60-tonne (66.1 ton) colossus mounted a 130mm (5.1in) B-13 L/50 naval gun in a very large box superstructure fitted to the experimental T-100 chassis.
SU-122	Soviet five-man diesel-powered 30.5 tonne (33.6 ton) self-propelled gun produced during 1943; it featured a potent 122mm (4.8in) L/23 field howitzer in an armoured sloped superstructure fitted on top of chassis of the T-34 tank.
KV-14	Prototype heavy assault gun design based upon the KV-1 tank chassis; with modification, entered production as the SU-152.
SU-152	Heavy diesel-powered assault gun that mounted the colossal 152mm (5.98in) L/29 field gun-howitzer in a thickly-armoured and well-sloped superstructure fitted to the chassis of the KV-1 tank; it weighed a considerable 45.5 tonnes (50.2 tons).
ISU-122	An interim low-silhouetted heavy assault gun design based on the chassis of the IS heavy tank family; mounted the 122mm (4.8in) A-19S gun.
ISU-122S	A 45.53 tonne (50.2 ton) variant of the ISU-122, manufactured from mid-1944, which featured the longer-barrelled and faster-firing 122mm (4.8in) D-25S gun.
ISU-152	A late-war four-man diesel-powered multi-role self-propelled gun that mounted the 152mm (5.98in) L/29 gun in a thickly-protected casement-style armoured superstructure fitted on top of the IS heavy tank chassis.

FOREIGN TANKS IN RED ARMY SERVICE

BRITISH

Infantry Tank Mk II Matilda II	A four-man heavily-protected 26.93 tonne (29.7 ton) infantry support tank that mounted a '2-pounder' 40mm (1.57in) OQF L/52 gun and featured 20–78mm (0.78–3.1in) armoured plates; it was powered by two mated 6-cylinder 65kW (87hp) AEC petrol engines.
Infantry Tank Mk II Matilda IIA	In this variant a 7.7mm (0.303in) BESA machine-gun replaced the co-axial Vickers mounted in the Matilda II.
Infantry Tank Mk II Matilda III	This variant of the Matilda incorporated two mated 71kW (95hp) Leyland engines instead of the AEC petrol engines.
Infantry Tank Mk II Matilda IV	Further up-rated Leyland engines powered this Matilda variant.
Infantry Tank Mk III Valentine II	This three-man, 17.3 tonne (19.1 ton), AEC diesel-powered medium infantry support tank mounted the '2-pounder' 40mm (1.57in) OQF L/52 gun in its welded turret and was protected by 8-65mm (0.32–2.56in) armoured plates.
Infantry Tank Mk III Valentine IV	This up-powered Valentine incorporated a 98kW (131hp) GMC-6004 diesel engine.
Infantry Tank Mk III Valentine V	In this model an improved American-designed transmission was incorporated into the vehicle.
Infantry Tank Mk III Valentine VII	A Canadian-built variant that included minor internal changes, including a new wireless radio set.
Infantry Tank Mk III Valentine [76.2mm]	The Soviets mounted their 76.2mm (3in) gun on a few Valentine II, IV, V, and VII tanks instead of the '2-pounder'.
Infantry Tank Mk III Valentine IX	This up-gunned variant mounted the more potent '6-pounder' 57mm (2.24in) OQF gun in a welded turret.
Infantry Tank Mk III Valentine X	The final variant employed by the Soviets, it incorporated a re-designed turret.
Infantry Tank Mk IV Churchill II	A heavily-armoured and relatively immobile heavy infantry support tank, the 39.9 tonne (43.8 ton) five-man Churchill II mounted the '2-pounder' 40mm (1.57in) OQF L/52 gun in a cuboid turret; its 16–102mm (0.63–4.02in) thick armour provided high levels of survivability.
Infantry Tank Mk IV Churchill III	This variant sported the more potent 57mm (2.24in) OQF gun in a modified large square-edged turret.
Infantry Tank Mk IV Churchill IV	This '6-pounder' gunned variant featured a re-designed round-edged turret.
Light Tank Mk VII Tetrarch	A small, fully-tracked, eight-wheeled light tank, this 7.6 tonne (8.4 ton) vehicle mounted a '2-pounder' 40mm (1.57in) OQF L/52 gun and could obtain an impressive maximum road speed of 64km/h (40mph).

UNITED STATES

M3 Stuart I	A small but relatively tall American-designed light tank, the 12.9 tonne (14.2 ton) M3 Stuart II mounted a 37mm (1.46in) M5 or M6 L/56.6 gun as well as four or five 7.62mm (0.3in) Browning M1919-A4 machine-guns; the vehicle was powered by a 186kW (250 hp) Continental W-670 petrol engine and employed the VVS Suspension (VVSS).
M3 Stuart II	This variant was similar to the Stuart I but instead featured a Guiberson T-1070 diesel engine.
M3A1 Stuart III	This improved version incorporated a re-designed rounded-edged turret and discontinued the two machine-guns located in the hull front sponsons.
M3 General Lee	The unusually-designed, tall-silhouetted, M3 General Lee medium tank sported a hull sponson-mounted 75mm (2.95in) M2/M3 howitzer, a 37mm (1.46in) M5 or M6 L/56.6 gun in its turret, and on the latter's roof an auxiliary turret with a machine-gun; this seven-man, 27.2 tonne (29.9 ton) vehicle was protected by 30–57mm (1.2–2.24in) armoured plates.
M4A2 Sherman (early production)	This five-man, 28.4 tonne (31.3 ton), multi-role medium tank mounted a 75mm (2.95in) M3 L/40 gun plus two Browning machine-guns; it sported 38–76mm (1.5–2.99in) armoured plates, and was powered by a 306kW (410hp) General Motors GM-6046 'twin' 2x 6-cylinder diesel engine, and had a more shallowly-sloped frontal plate.
M4A2 Sherman (late production)	This up-armoured late production variant, with steeper-sloped frontal plate, had its maximum armoured protection increased to a thickness of 108mm (4.25in), which contributed to the vehicle's weight increasing to 31.8 tonnes (35.1 tons).
M4A2 Sherman 76(W)	This Sherman variant featured a new T-23-style turret that mounted the longer-barrelled 76.2mm (3in) M1A1 L/52 gun, for which it stowed 71 rounds, housed in 'Wet' conditions to minimize the risk of fire when hit.
M4A4 Sherman	Small numbers of this petrol-powered Sherman variant served with the Red Army; it was powered by a 350kW (470hp) Chrysler A-57 multi-bank 30-cylinder petrol engine.
M10 Wolverine (early production)	This 29.13 tonne (32.1 ton) 'Three-inch Gun Motor Carriage' [GMC] turreted tank destroyer was based upon the M4A2 Sherman chassis; its angular five-sided open-roofed turret mounted the 76.2mm (3in) M7 L/50 gun and the vehicle was powered by a 306kW (410hp) General Motors GM-6046 6-71 'twin' 2x 6-cylinder diesel engine.
M10 Wolverine (mid-production)	This tank destroyer variant had its turret projected backwards to form a heavy triangular counter-weight to balance the gun's weight; like its predecessor, it was protected by 9.5–57.2mm (0.8–2.25in) armour.
M10 Wolverine (late production)	In these late production batches, the tank destroyer mounted the 76.2mm (3in) M1 gun, rather than the M7, in a counter-weighted turret.
M10A1 Wolverine	This tank destroyer variant with counter-weighted turret was based on the chassis of the M4A3 Sherman; it was powered by a 336kW (450hp) Ford GAA diesel engine.

Tank Profiles Compared

For ease of identification and comparison, below are the main tanks to serve in the Red Army during World War II in profile form.

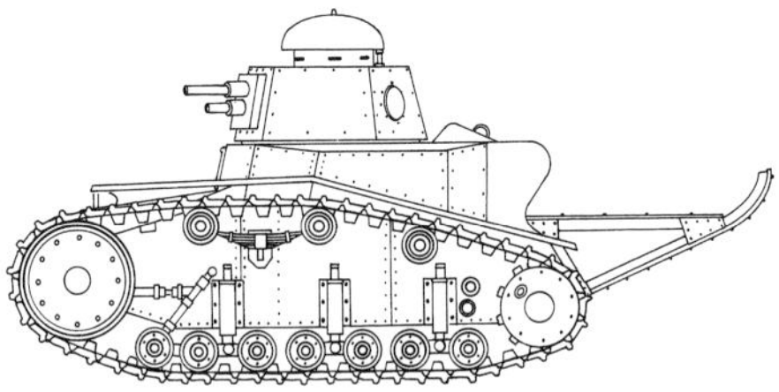

MS-3 (T-18)

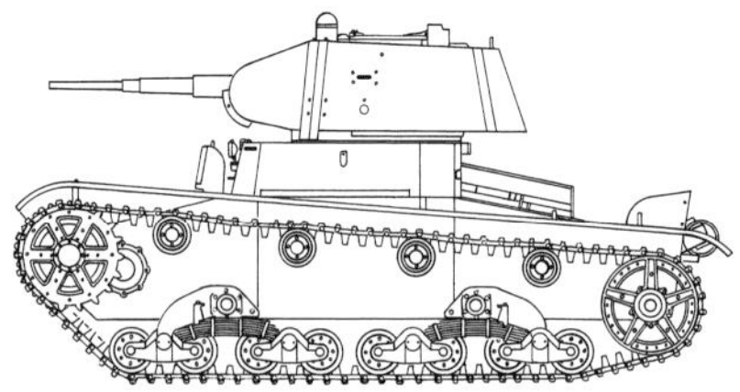

T-26 Model 1939

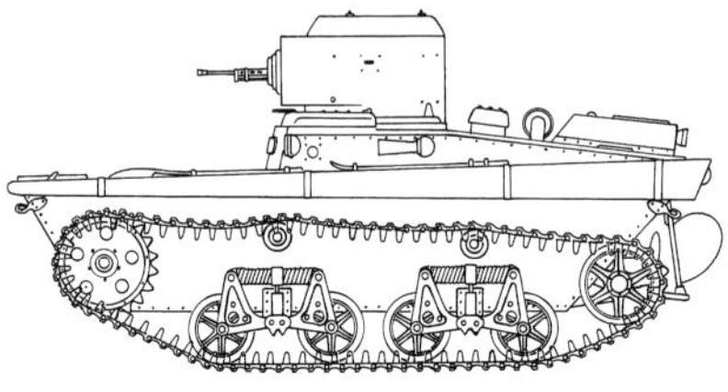

T-37A

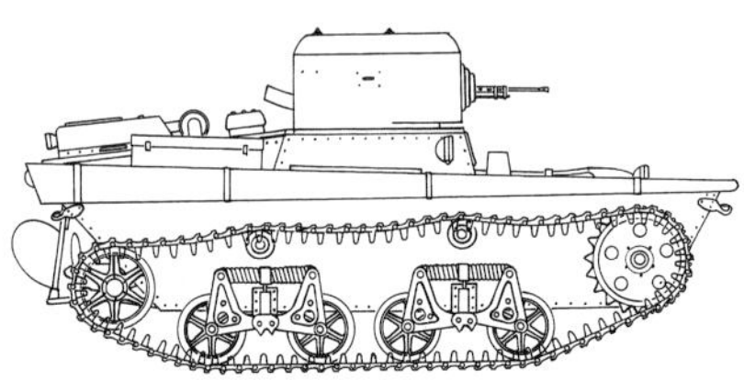

T-38

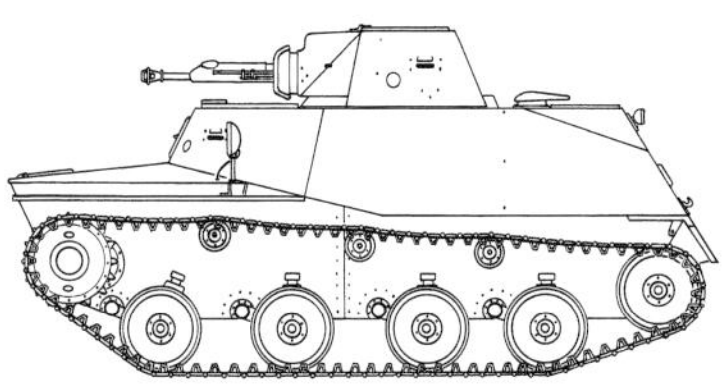

T-40

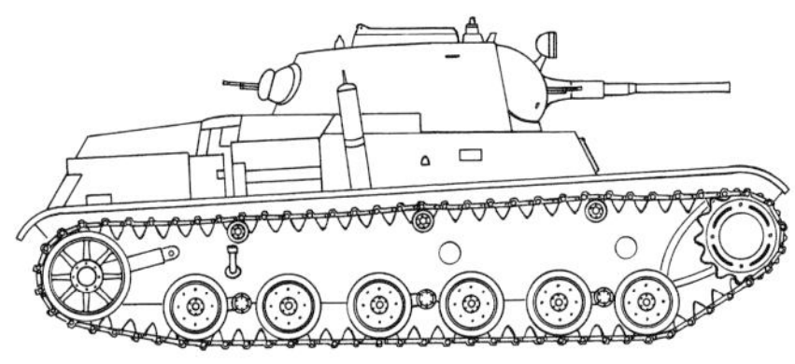

T-46

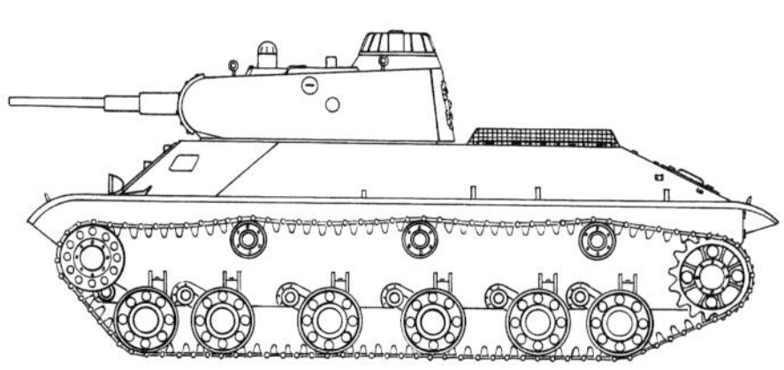

T-50

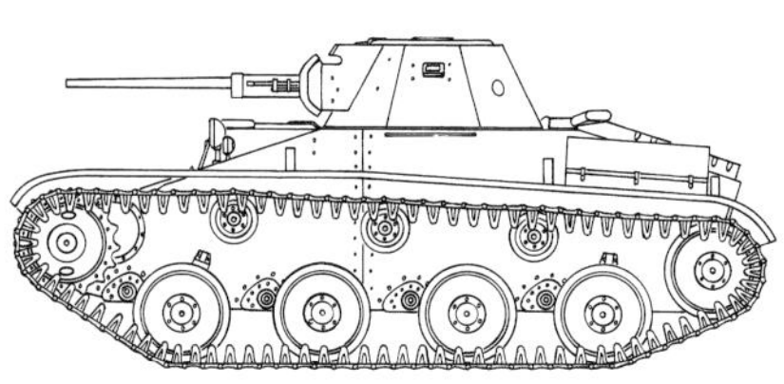

T-60

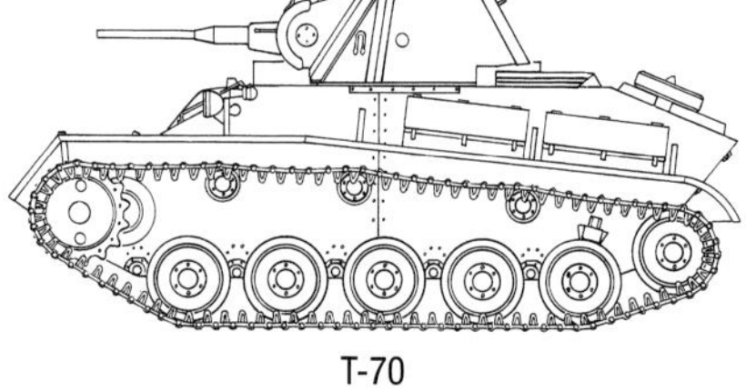

T-70

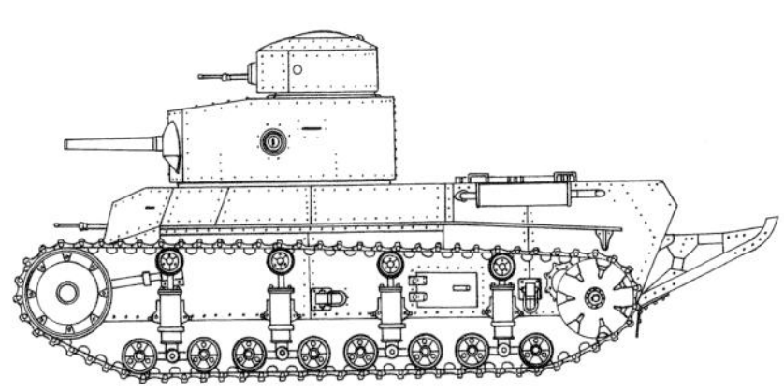

T-24

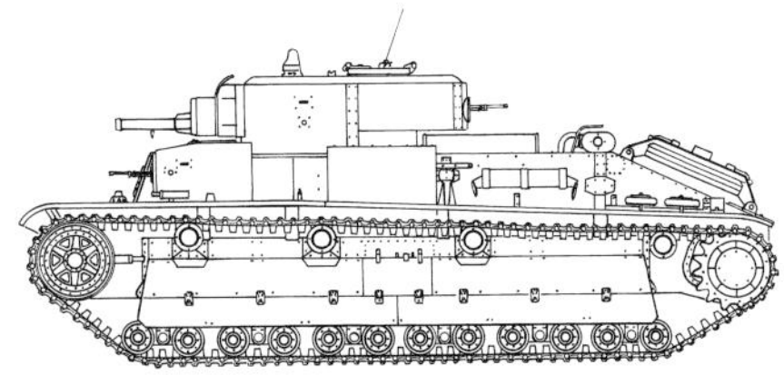

T-28

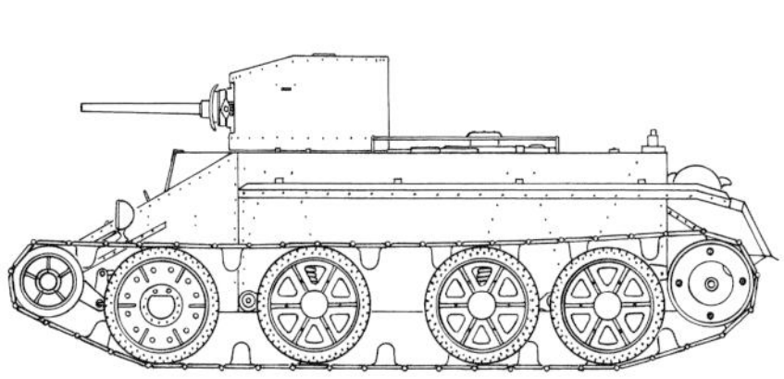

BT-2

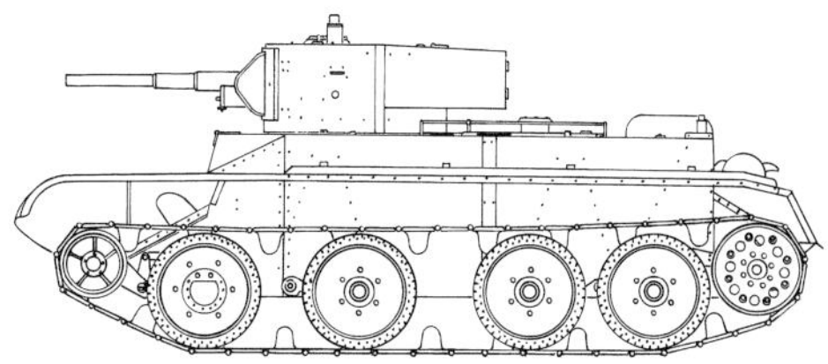

BT-5

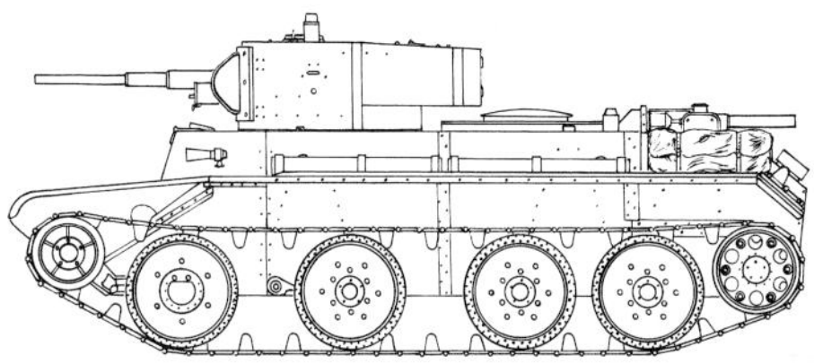

BT-7

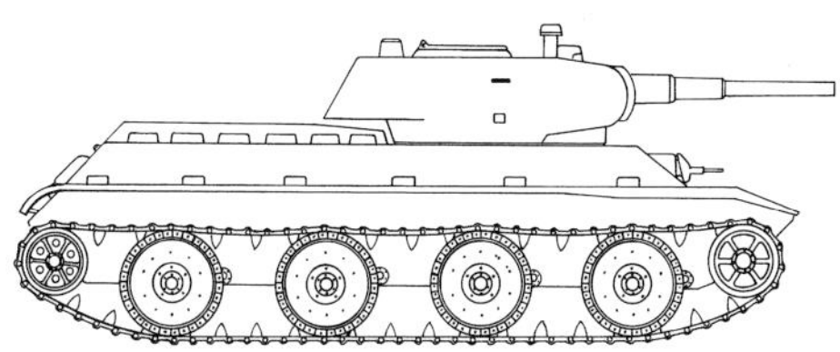

A-20

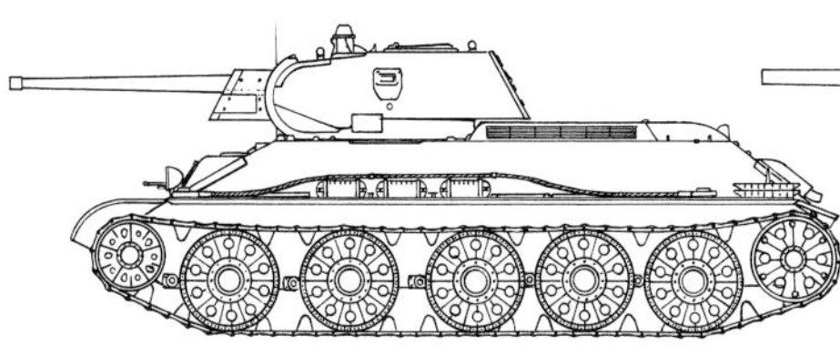

T-34/76 Model 1942

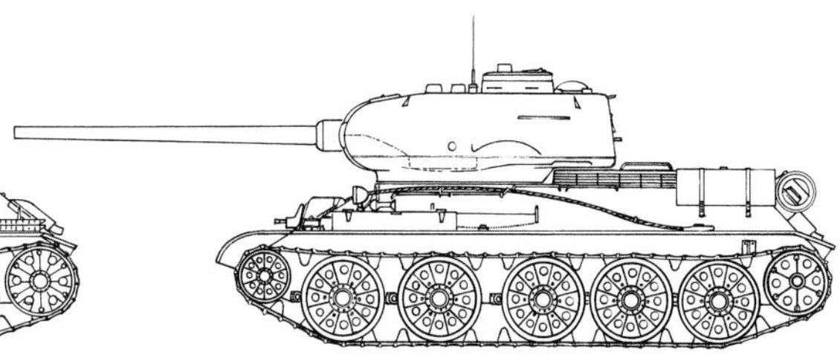

T-34/85

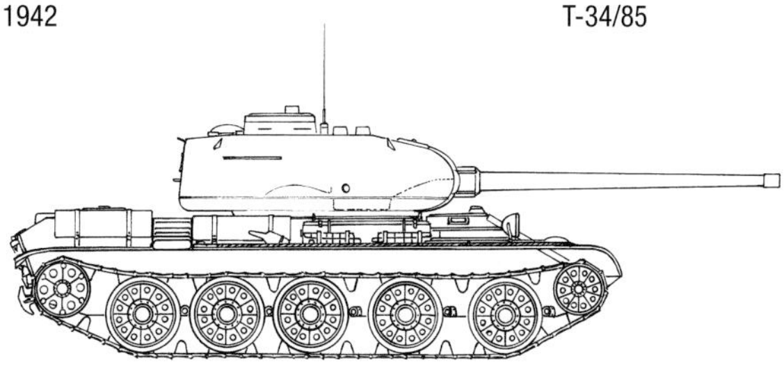

T-44

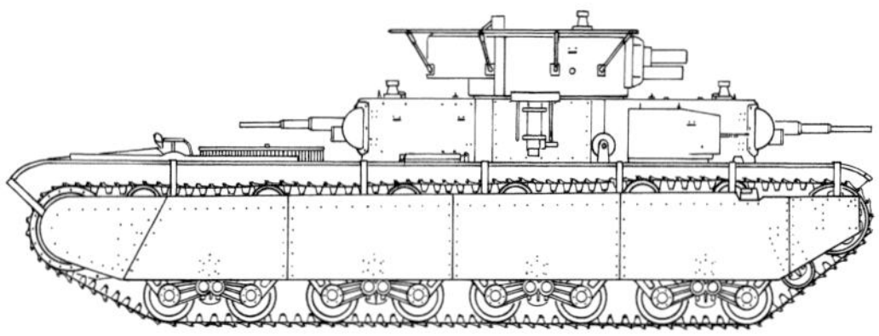

T-35

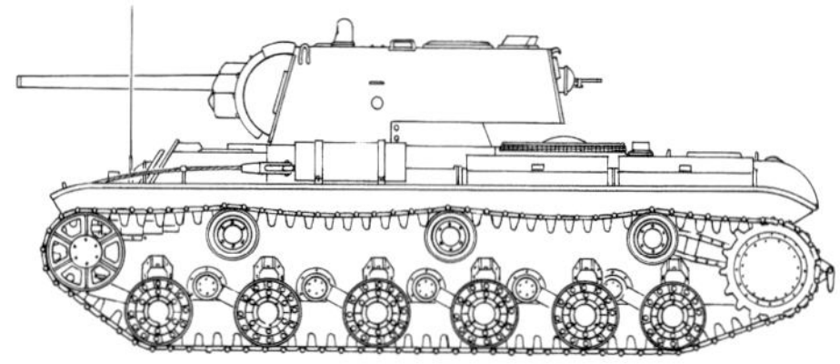

KV-1 Model 1941

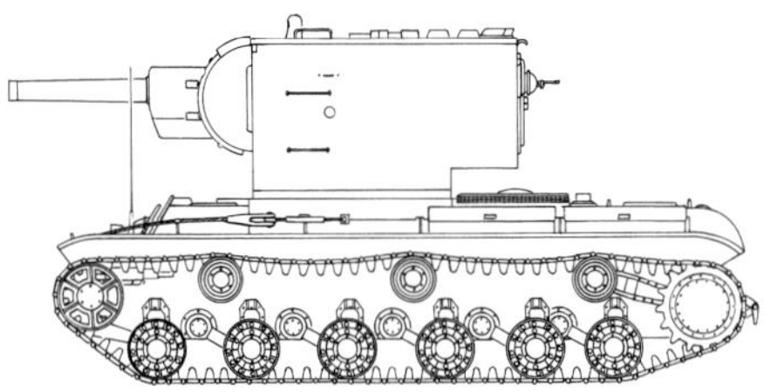

KV-2 Model 1940

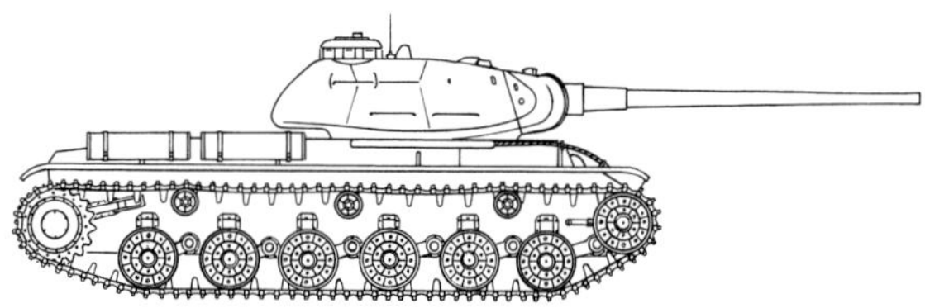

KV-85

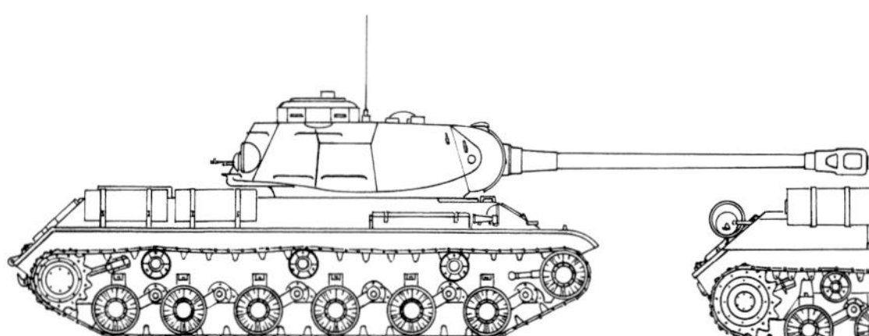

IS-2M

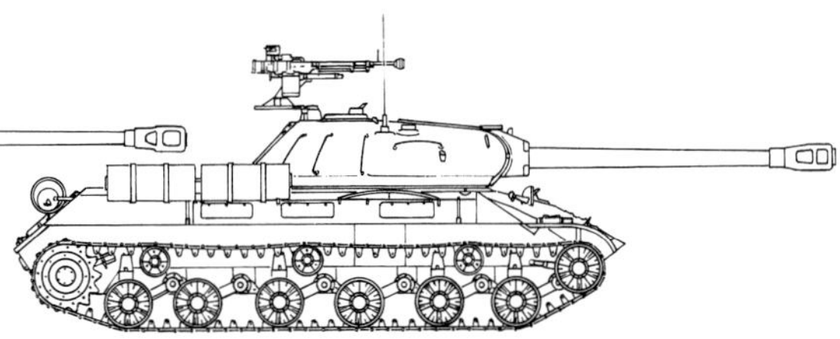

IS-3

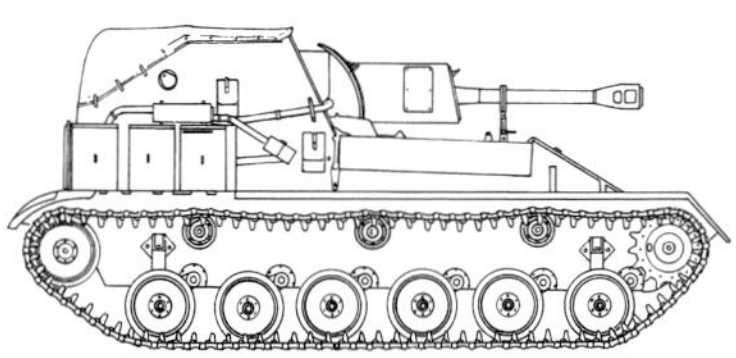

SU-76M

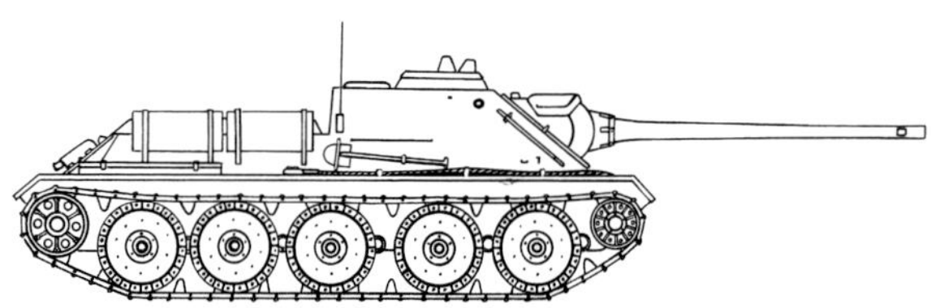

SU-85

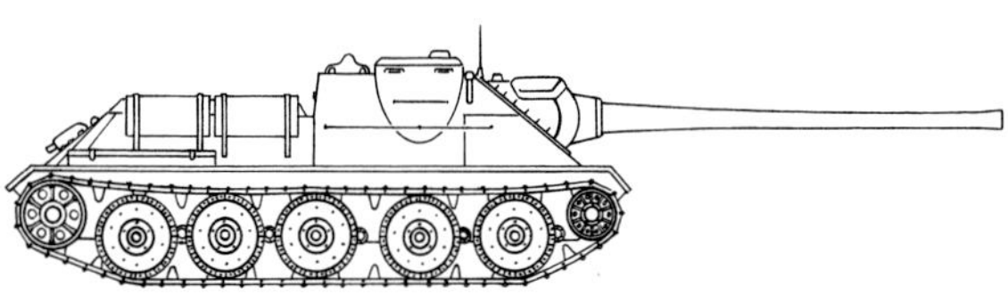

SU-100

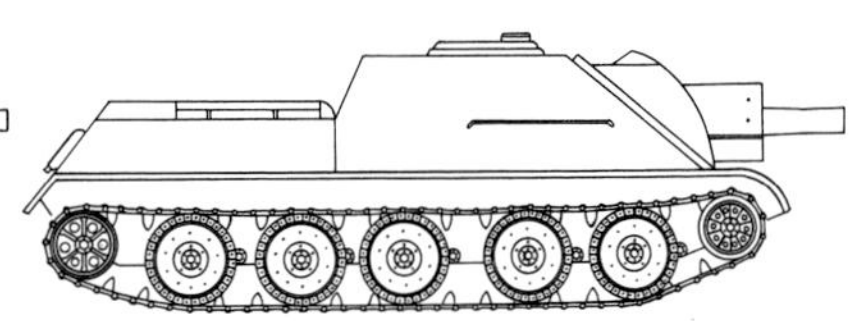

SU-122

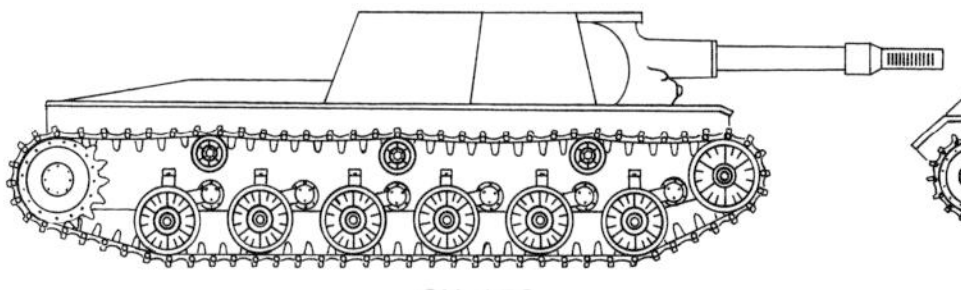

SU-152

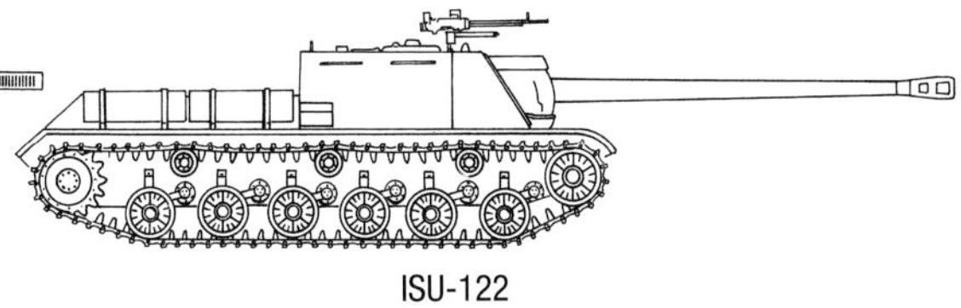

ISU-122

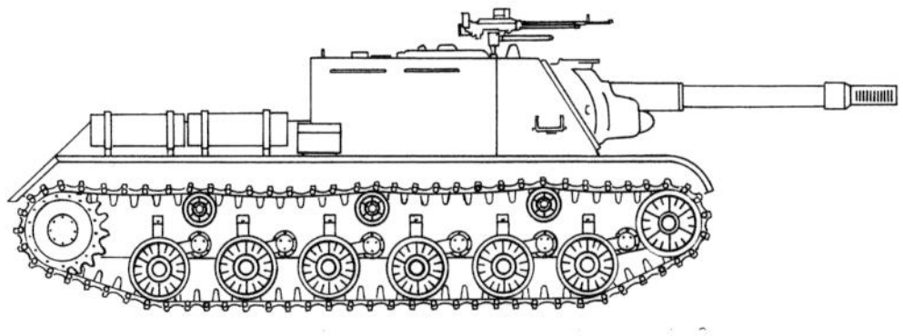

ISU-152

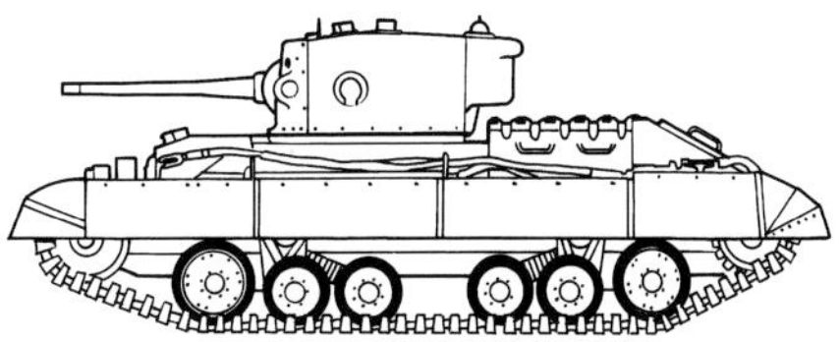

Valentine II

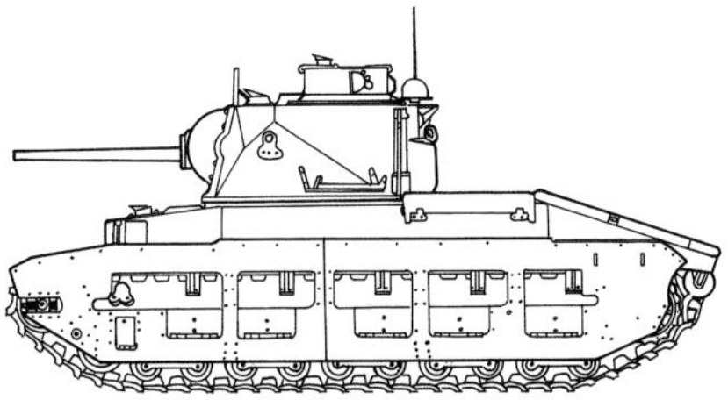

Matilda II

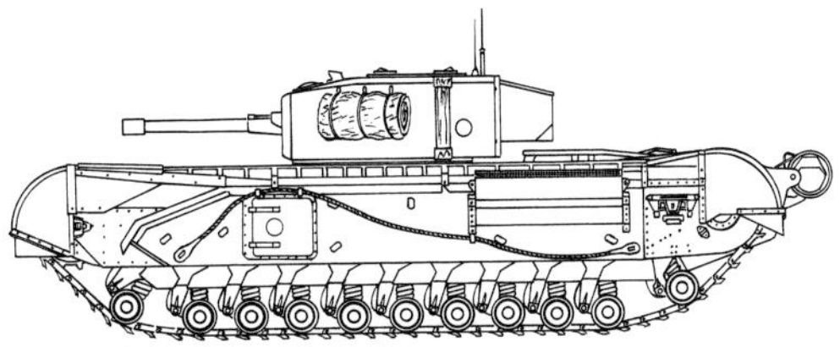

Churchill III

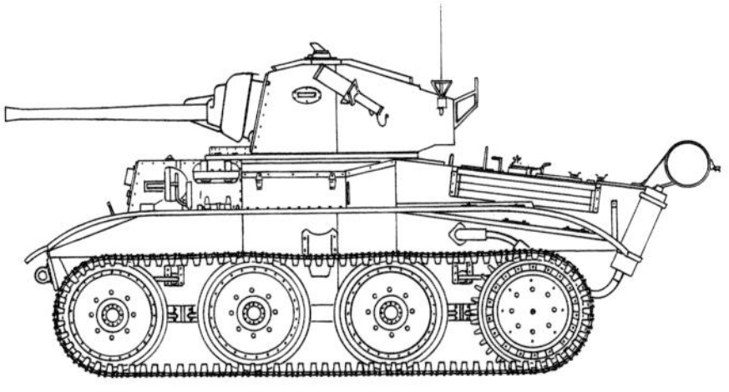

Tetrarch

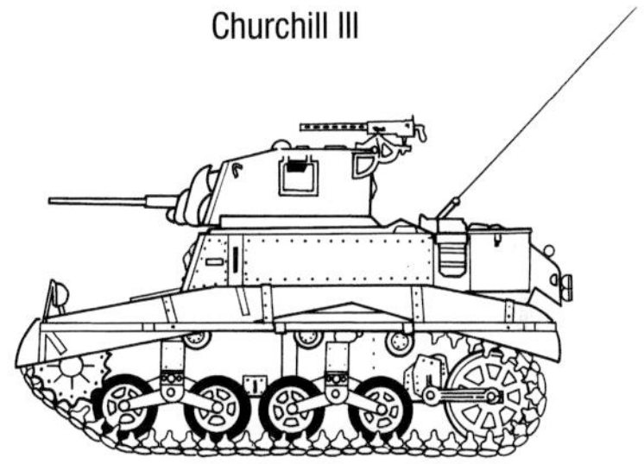

M3A1 Stuart

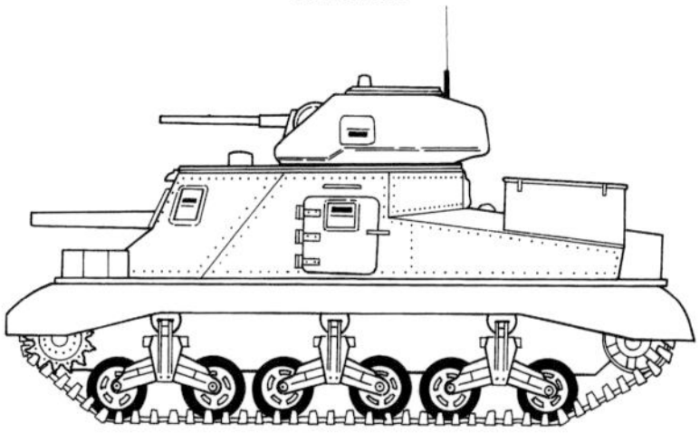

M3 Lee

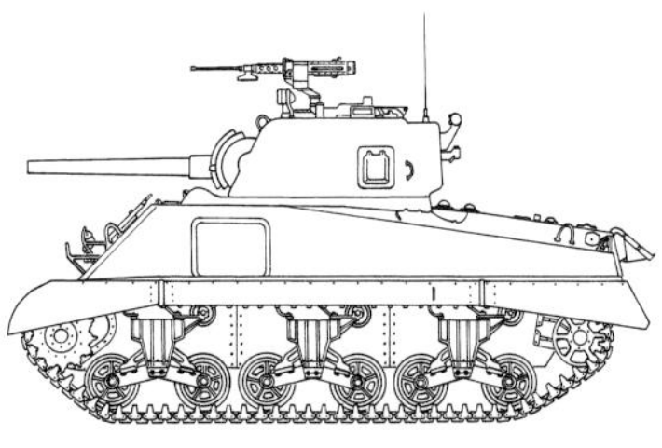

M4A2 Sherman

Soviet-produced Tanks: Specifications Compared

Below is a series of tables divided by type that compares the main variants of Soviet-made tanks to serve with the Red Army during World War II.

Red Army soldiers ride aboard a T-34/76 through western Russia during the summer offensive of 1944. For its combination of speed, durability, armament and ease of manufacture, many consider the T-34 to be the most significant tank of World War II.

Light Tanks	T-18	T-18M	T-27	T-26 Model 1931
Date	1927	1938	1931–33	1932–33
Number produced	959		4258	1261
Crew	2	2	2	3
Weight (tonnes)	5.3	5.95	2.7	8.2
Length (m)	4.38	4.38	2.60	4.90
Width (m)	1.76	1.76	1.83	3.40
Height (m)	2.10	2.10	1.44	2.36
Armament	37mm Hochkiss 6.5mm Fyodorov MG	37mm PS-1 7.62mm DT MG	1 x 7.62 mm DT MG	2 x 7.62 DT MG
Engine	35hp	GAZ M-1, 50hp	GAZ-AA, 40hp	T-26, 90hp
Max road speed (km/h)	17	24	42	35
Armour (mm)				
turret	9	9	–	13–15
hull front	14	14	10	13–15
hull side	14	14	10	13–15
hull rear	–	–	10	13–15
hull roof	–	–	6	6
hull floor	–	–	4	6

Light Tanks	KhT-26		T-26 Model 1933		T-26 Model 1934	
Date	1933–34		1934		1935	
Number produced	552		5584 (all models)			
Crew	2		3		3	
Weight (tonnes)	9.8		9.4		9.6	
Length (m)	4.90		4.90		4.90	
Width (m)	3.40		3.40		3.40	
Height (m)	2.36		2.31		2.31	
Armament	KS-24	1 x 7.62mm DT MG	45mm 20K	1 x 7.62mm DT MG	45mm 20K	1 x 7.62mm DT MG
Engine	T-26, 90hp		T-26, 90hp		T-26, 90hp	
Max road speed (km/h)	35		35		35	
Armour (mm)						
turret	15		15		15	
hull front	15		15		15	
hull side	15		15		15	
hull rear	15		15		15	
hull roof	6		6		6–10	
hull floor	6		6		6	

Light Tanks	T-26 Model 1936		T-26 Model 1938		KhT-130	
Date	1936		1937–38		1938–39	
Number produced	5584 (all models)				401	
Crew	3		3		3	
Weight (tonnes)	9.65		9.8		10	
Length (m)	4.90		4.90		4.90	
Width (m)	3.40		3.40		3.40	
Height (m)	2.31		2.31		2.31	
Armament	45mm 20K	3 x 7.62mm DT MG	45mm 20K	2 x 7.62mm DT MG	KS-25	1 x 7.62mm DT MG
Engine	T-26, 90hp		T-26, 95hp		T-26, 95hp	
Max road speed (km/h)	35		35		35	
Armour (mm)						
turret	15		15		15	
hull front	15		15		15	
hull side	15		15		15	
hull rear	15		15		15	
hull roof	6–10		6–10		6–10	
hull floor	6		6		6	

Light Tanks	T-26 Model 1939		KhT-133		T-50	
Date	1939–40		1940		1941	
Number produced	2678		269		69	
Crew	3		3		4	
Weight (tonnes)	10.25		10.5		13.7	
Length (m)	4.90		4.90		5.20	
Width (m)	3.40		3.40		2.47	
Height (m)	2.31		2.31		2.16	
Armament	45mm 20K	2 x 7.62mm DT MG	KS-25	2 x 7.62mm DT MG	45mm Model 1938	1 x 7.62mm DT MG
Engine	T-26, 95hp		T-26, 95hp		V-4, 300hp	
Max road speed (km/h)	34		35		60	
Armour (mm)						
turret	15–20		15–20		37	
hull front	15		15		37	
hull side	15		15		37	
hull rear	15–20		15–20		25	
hull roof	6–10		6–10		15	
hull floor	6		6		12	

Light Tanks	T-60		T-70		T-80	
Date	1942–43		1942–43		1943	
Number produced	6292		8225		81	
Crew	2		2		3	
Weight (tonnes)	5.2		12.67		11.6	
Length (m)	4.29		4.29		4.29	
Width (m)	2.46		2.46		2.46	
Height (m)	1.89		1.89		1.89	
Armament	20mm TNSh	1 x 7.62mm DT MG	45mm 20K	1 x 7.62mm DT MG	45mm 20K	1 x 7.62mm DT MG
Engine	GAZ-202, 70hp		2 x GAZ-70-6000, 2 x 70hp		2 x GAZ-11, 2 x 85hp	
Max road speed (km/h)	45		45		47	
Armour (mm)						
turret	25		45		35–50	
hull front	35		35–45		35–45	
hull side	15		15		25	
hull rear	25		35		35	
hull roof	10		10		10–15	
hull floor	10		10		10	

Fast Tanks	BT-2		BT-2	BT-5 Model 1933	
Date	1932		1933	1933–34	
Number produced	208		412	1884	
Crew	3		3	3	
Weight (tonnes)	10.2		10.2	11.5	
Length (m)	5.58		5.58	5.58	
Width (m)	2.30		2.30	2.30	
Height (m)	2.25		2.25	2.25	
Armament	37mm B-3	1 x 7.62mm DT MG	2–3 x 7.62mm DT MG	45mm 20K	1 x 7.62mm DT MG
Engine	M5-400, 400hp		M5-400, 400hp	M5-400, 400hp	
Max road speed (km/h)	100		100	72	
Armour (mm)					
turret	13		13	13	
hull front	13		13	13	
hull side	13		13	13	
hull rear	13		13	13	
hull roof	–		–	–	
hull floor	–		–	–	

Fast Tanks	BT-7		BT-7A		BT-7M (BT-8)	
Date	1935–38		1936–38		1938–40	
Number produced	4540		154		790	
Crew	3		3		3	
Weight (tonnes)	14		14.5		14.7	
Length (m)	5.66		5.66		5.66	
Width (m)	2.29		2.29		2.29	
Height (m)	2.42		2.42		2.52	
Armament	45mm 20K	1 x 7.62mm DT MG	76.2mm KT	2 x 7.62mm DT MG	45mm 20K	3 x 7.62mm DT MG
Engine	M-17T, 500hp		V-2, 500hp		M-17T, 450hp	
Max road speed (km/h)	86		86		86	
Armour (mm)						
turret	13–15		13–15		13–15	
hull front	15		15		20–22	
hull side	15		15		15	
hull rear	13		13		13	
hull roof	10		10		10	
hull floor	6		6		6	

<table>
<tr><th>Amphibious Tanks</th><th>T-37A</th><th>T-38</th><th colspan="2">T-40</th><th colspan="2">T-30 (T-40S)</th></tr>
<tr><td>Date</td><td>1933–36</td><td>1936–39</td><td colspan="2">1940</td><td colspan="2">1941</td></tr>
<tr><td>Number produced</td><td>2627</td><td>1382</td><td colspan="4">709</td></tr>
<tr><td>Crew</td><td>2</td><td>2</td><td colspan="2">2</td><td colspan="2">2</td></tr>
<tr><td>Weight (tonnes)</td><td>3.2</td><td>3.3</td><td colspan="2">5.5</td><td colspan="2">5.6</td></tr>
<tr><td>Length (m)</td><td>3.75</td><td>3.76</td><td colspan="2">4.43</td><td colspan="2">4.43</td></tr>
<tr><td>Width (m)</td><td>2.1</td><td>1.95</td><td colspan="2">2.51</td><td colspan="2">2.51</td></tr>
<tr><td>Height (m)</td><td>1.82</td><td>1.62</td><td colspan="2">2.12</td><td colspan="2">2.12</td></tr>
<tr><td>Armament</td><td>1 x 7.62mm DT MG</td><td>1 x 7.62mm DT MG</td><td>12.7mm DShK</td><td>1 x 7.62mm DT MG</td><td>20mm TNSh</td><td>1 x 7.62mm DT MG</td></tr>
<tr><td>Engine</td><td>GAZ-AA, 40hp</td><td>GAZ-AA, 40hp</td><td colspan="2">GAZ-11, 85hp</td><td colspan="2">GAZ-11, 85hp</td></tr>
<tr><td>Max road speed (km/h)</td><td>35</td><td>40</td><td colspan="2">45</td><td colspan="2">45</td></tr>
<tr><td>Armour (mm)</td><td></td><td></td><td colspan="2"></td><td colspan="2"></td></tr>
<tr><td>turret</td><td>8</td><td>8</td><td colspan="2">13</td><td colspan="2">20</td></tr>
<tr><td>hull front</td><td>9</td><td>9</td><td colspan="2">13</td><td colspan="2">15</td></tr>
<tr><td>hull side</td><td>9</td><td>9</td><td colspan="2">13</td><td colspan="2">15</td></tr>
<tr><td>hull rear</td><td>–</td><td>9</td><td colspan="2">13</td><td colspan="2">13</td></tr>
<tr><td>hull roof</td><td>–</td><td>4</td><td colspan="2">5</td><td colspan="2">5</td></tr>
<tr><td>hull floor</td><td>–</td><td>4</td><td colspan="2">5</td><td colspan="2">5</td></tr>
</table>

<table>
<tr><th>Multi-turreted Tanks</th><th colspan="2">T-28 Model 1934</th><th colspan="2">T-28 Model 1938</th><th colspan="3">T-35</th></tr>
<tr><td>Date</td><td colspan="2">1933–37</td><td colspan="2">1940</td><td colspan="3">1933–39</td></tr>
<tr><td>Number produced</td><td colspan="2">392</td><td colspan="2">111</td><td colspan="3">61</td></tr>
<tr><td>Crew</td><td colspan="2">6</td><td colspan="2">6</td><td colspan="3">11</td></tr>
<tr><td>Weight (tonnes)</td><td colspan="2">28</td><td colspan="2">32</td><td colspan="3">50</td></tr>
<tr><td>Length (m)</td><td colspan="2">7.44</td><td colspan="2">7.44</td><td colspan="3">9.72</td></tr>
<tr><td>Width (m)</td><td colspan="2">2.81</td><td colspan="2">2.81</td><td colspan="3">3.20</td></tr>
<tr><td>Height (m)</td><td colspan="2">2.82</td><td colspan="2">2.82</td><td colspan="3">3.43</td></tr>
<tr><td>Armament</td><td>76.2mm KT</td><td>3 x 7.62mm DT MG</td><td>76.2mm L-10</td><td>5 x 7.62mm DT MG</td><td>476.2mm KT</td><td>2 x 45mm 20K</td><td>6 x 7.62mm DT MG</td></tr>
<tr><td>Engine</td><td colspan="2">M-17T, 500hp</td><td colspan="2">M-17T, 500hp</td><td colspan="3">M-17T, 500hp</td></tr>
<tr><td>Max road speed (km/h)</td><td colspan="2">37</td><td colspan="2">37</td><td colspan="3">30</td></tr>
<tr><td>Armour (mm)</td><td colspan="2"></td><td colspan="2"></td><td colspan="3"></td></tr>
<tr><td>turret</td><td colspan="2">20</td><td colspan="2">50</td><td colspan="3">20–30</td></tr>
<tr><td>hull front</td><td colspan="2">30</td><td colspan="2">50</td><td colspan="3">30</td></tr>
<tr><td>hull side</td><td colspan="2">20</td><td colspan="2">50</td><td colspan="3">20</td></tr>
<tr><td>hull rear</td><td colspan="2">20</td><td colspan="2">50</td><td colspan="3">20</td></tr>
<tr><td>hull roof</td><td colspan="2">15</td><td colspan="2">50</td><td colspan="3">10–15</td></tr>
<tr><td>hull floor</td><td colspan="2">15</td><td colspan="2">15</td><td colspan="3">20</td></tr>
</table>

<table>
<tr><th>Medium Tanks</th><th colspan="2">T-34 Model 1940</th><th colspan="2">T-34 Model 1941</th><th colspan="2">T-34 Model 1942</th></tr>
<tr><td>Date</td><td colspan="2">1940</td><td colspan="2">1941–42</td><td colspan="2">1942</td></tr>
<tr><td>Number produced</td><td colspan="6">33,805 (all models)</td></tr>
<tr><td>Crew</td><td colspan="2">4</td><td colspan="2">4</td><td colspan="2">4</td></tr>
<tr><td>Weight (tonnes)</td><td colspan="2">26.8</td><td colspan="2">28.5</td><td colspan="2">28.5</td></tr>
<tr><td>Length (m)</td><td colspan="2">5.92</td><td colspan="2">5.92</td><td colspan="2">5.92</td></tr>
<tr><td>Width (m)</td><td colspan="2">3</td><td colspan="2">3</td><td colspan="2">3</td></tr>
<tr><td>Height (m)</td><td colspan="2">2.45</td><td colspan="2">2.45</td><td colspan="2">2.45</td></tr>
<tr><td>Armament</td><td>76.2mm L-11</td><td>2 x 7.62mm DT MG</td><td>76.2mm F-34</td><td>2 x 7.62mm DT MG</td><td>7.62mm F-34</td><td>2 x 7.62mm DT MG</td></tr>
<tr><td>Engine</td><td colspan="2">V-2-34, 500hp</td><td colspan="2">V-2-34, 500hp</td><td colspan="2">V-2-34, 500hp</td></tr>
<tr><td>Max road speed (km/h)</td><td colspan="2">55</td><td colspan="2">55</td><td colspan="2">55</td></tr>
<tr><td>Armour (mm)</td><td colspan="2"></td><td colspan="2"></td><td colspan="2"></td></tr>
<tr><td>turret</td><td colspan="2">45</td><td colspan="2">52–60</td><td colspan="2">45–52</td></tr>
<tr><td>hull front</td><td colspan="2">45</td><td colspan="2">45</td><td colspan="2">45</td></tr>
<tr><td>hull side</td><td colspan="2">45</td><td colspan="2">45</td><td colspan="2">45</td></tr>
<tr><td>hull rear</td><td colspan="2">40</td><td colspan="2">40</td><td colspan="2">40</td></tr>
<tr><td>hull roof</td><td colspan="2">16–20</td><td colspan="2">20</td><td colspan="2">20</td></tr>
<tr><td>hull floor</td><td colspan="2">16</td><td colspan="2">20</td><td colspan="2">20</td></tr>
</table>

<table>
<tr><th>Medium Tanks</th><th colspan="2">T-34 Model 1943</th><th colspan="2">1943 T-34-57</th><th colspan="3">OT-34</th></tr>
<tr><td>Date</td><td colspan="2">1943–44</td><td colspan="2">1941–43</td><td colspan="3">1943</td></tr>
<tr><td>Number produced</td><td colspan="4">33,805</td><td colspan="3">1170</td></tr>
<tr><td>Crew</td><td colspan="2">4</td><td colspan="2">4</td><td colspan="3">4</td></tr>
<tr><td>Weight (tonnes)</td><td colspan="2">30.9</td><td colspan="2">30.9</td><td colspan="3">31</td></tr>
<tr><td>Length (m)</td><td colspan="2">5.92</td><td colspan="2">5.92</td><td colspan="3">5.92</td></tr>
<tr><td>Width (m)</td><td colspan="2">3</td><td colspan="2">3</td><td colspan="3">3</td></tr>
<tr><td>Height (m)</td><td colspan="2">2.45</td><td colspan="2">2.45</td><td colspan="3">2.45</td></tr>
<tr><td>Armament</td><td>76.2mm F-34</td><td>2 x 7.62mm DT MG</td><td>57mm ZIS-4</td><td>2 x 7.62mm DT MG</td><td>ATO-41</td><td>76.2mm F-34</td><td>1 x 7.62mm DT MG</td></tr>
<tr><td>Engine</td><td colspan="2">V-2-34, 500hp</td><td colspan="2">V-2-34, 500hp</td><td colspan="3">V-2-34, 500hp</td></tr>
<tr><td>Max road speed (km/h)</td><td colspan="2">55</td><td colspan="2">55</td><td colspan="3">55</td></tr>
<tr><td>Armour (mm)</td><td colspan="2"></td><td colspan="2"></td><td colspan="3"></td></tr>
<tr><td>turret</td><td colspan="2">45–52</td><td colspan="2">45–52</td><td colspan="3">45–52</td></tr>
<tr><td>hull front</td><td colspan="2">45</td><td colspan="2">45</td><td colspan="3">45</td></tr>
<tr><td>hull side</td><td colspan="2">45</td><td colspan="2">45</td><td colspan="3">45</td></tr>
<tr><td>hull rear</td><td colspan="2">45</td><td colspan="2">45</td><td colspan="3">45</td></tr>
<tr><td>hull roof</td><td colspan="2">20</td><td colspan="2">20</td><td colspan="3">20</td></tr>
<tr><td>hull floor</td><td colspan="2">20</td><td colspan="2">20</td><td colspan="3">20</td></tr>
</table>

The crew of a T-34/76 Model 1943 huddle on the tank's turret during the fighting around Stalingrad as part of Operation Uranus, the Red Army's counteroffensive in the winter of 1942/43.

Medium Tanks	T-34-85 Model 1943	T-34-85 Model 1944	T-44
Date	1944	1944–47	1944–45
Number produced	300+	22,278	175
Crew	4	5	5
Weight (tonnes)	32	32	31.9
Length (m)	6.1	6.1	7.65
Width (m)	3	3	3.15
Height (m)	2.72	2.72	2.45
Armament	85mm D-5T 2 x 7.62mm DT MG	85mm S-53 or ZIS-S-53 2 x 7.62mm DT MG	85mm D-5T or ZIS-S-53 2 x 7.62mm DT MG
Engine	V-2-34, 500hp	V-2-34, 500hp	V-44, 500hp
Max road speed (km/h)	55	55	51
Armour (mm)			
turret	52–90	52–90	52–120
hull front	60	60	90
hull side	45	45	45
hull rear	45	45	45
hull roof	20	20	20
hull floor	20	20	20

Heavy Tanks	**KV-2**		**KV-1 Model 1940**		**KV-1 Model 1941**	
Date	1940–41		1940–41		1941	
Number produced	334		2955 (all models)			
Crew	6		5		5	
Weight (tonnes)	52		43.6		45	
Length (m)	7.31		6.68		6.75	
Width (m)	3.49		3.32		3.32	
Height (m)	3.93		2.71		2.71	
Armament	152.4mm M-10T	3 x 7.62mm DT MG	76.2mm L-11	3 x 7.62mm DT MG	76.2mm F-34	3 x 7.62mm DT MG
Engine	V-2K, 600hp		V-2K, 600hp		V-2K, 600hp	
Max road speed (km/h)	27		35		35	
Armour (mm)						
turret	75–110		75–90		90–110	
hull front	75		75		75	
hull side	75		75		75	
hull rear	60		60		60	
hull roof	30		30		30	
hull floor	30		30		30	

Heavy Tanks	**KV-1 Model 1942**		**KV-1S**		**KV-85**	
Date	1942		1942–43		1942	
Number produced	–		1370		148	
Crew	5		5		4	
Weight (tonnes)	48		42.5		46	
Length (m)	6.8		7.41		8.6	
Width (m)	3.32		2.87		3.25	
Height (m)	2.71		3.25		2.9	
Armament	76.2mm F-34	3 x 7.62mm DT MG	76.2mm ZIS-5	3 x 7.62mm DT MG	85mm D-5T	3 x 7.62mm DT MG
Engine	V-2K, 600hp		V-2K, 600hp		V-2K, 600hp	
Max road speed (km/h)	28		43		35	
Armour (mm)						
turret	75–120		75–82		110	
hull front	75–90		50–75		40–75	
hull side	75		60		60	
hull rear	60–70		60		40–75	
hull roof	30–40		30–40		30–40	
hull floor	30		30		20	

Heavy Tanks	IS-1		IS-2		IS-3		
Date	1943		1944–45		1945		
Number produced	107		3854		350		
Crew	4		4		4		
Weight (tonnes)	43.4		46		46.5		
Length (m)	8.56		9.9		9.85		
Width (m)	3		3.09		3.09		
Height (m)	2.74		2.73		2.45		
Armament	85mm D-5T	3 x 7.62mm DT MG	122mm D-25T	2 x 7.62mm DT MG	122mm D-25T	1 x 12.7mm DShK MG	1 x 7.62mm DT MG
Engine	V-2-IS, 600hp		V-2-IS, 600hp		V-2-IS, 600hp		
Max road speed (km/h)	37		37		40		
Armour (mm)							
turret	60–120		100–120		175		
hull front	100		90–100		110		
hull side	90		90		90		
hull rear	60		60		60		
hull roof	30		30		30		
hull floor	20		20		30		

Self-propelled Guns	ZiS-30		SU-76	SU-76M	SU-122
Date	1941		1943	1943–44	1942–43
Number produced	101		579	13,932	1150
Crew	4		3	4	5
Weight (tonnes)	4		10.3	10.2	30.5
Length (m)	3.45		5	5	6.95
Width (m)	1.86		2.7	2.7	3
Height (m)	2		2.1	2.1	2.32
Armament	57mm ZIS-2	7.62mm DT MG	76.2mm ZIS-3	76.2mm ZIS-3	122mm M-30
Engine	GAZ M-1, 50hp		GAZ-202, 2 x 70hp	GAZ-203, 2 x 70hp	V-2-34, 500hp
Max road speed (km/h)	40		45	45	55
Armour (mm)					
hull front	7–10		15–35	25–30	45
hull side	7–10		15–35	10–15	45
hull rear	7–10		15–35	0–15	45
hull roof	–		–	–	20
hull floor	–		7	7	20

Self-propelled Guns	SU-85	SU-85M	SU-100	SU-152	ISU-152
Date	1943–44	1944	1944–47	1943	1944–45
Number produced	2050		2350	704	1885
Crew	4	4	4	5	5
Weight (tonnes)	29.6	31.2	31.6	45.5	47.3
Length (m)	8.15	8.15	9.45	8.95	9.18
Width (m)	3	3	3	3.35	3.07
Height (m)	2.45	2.45	2.25	2.45	2.48
Armament	85mm D-5T	85mm D-5T	100mm D-10S	152.4mm ML-20S	152.4mm ML-20S
Engine	V-2-34, 500hp	V-2-34M, 520hp	V-2-34M, 520hp	V-2K, 600hp	V-2-IS, 520hp
Max road speed (km/h)	55	50	48	43	30
Armour (mm)					
hull front	45	75	75	60–75	60–90
hull side	45	45	45	60	75–90
hull rear	45	45	45	60	60
hull roof	20	20	20	20	30
hull floor	20	20	20	20	20

Self-propelled Guns	ISU-122	ISU-122S	
Date	1944	1944–45	
Number produced	2410		
Crew	5	5	
Weight (tonnes)	45.5	45.53	
Length (m)	9.84	9.84	
Width (m)	3.07	3.07	
Height (m)	2.84	2.84	
Armament	122mm A-19S	122mm D-25S	12.7mm DShK MG
Engine	V-2-IS, 520hp	V-2-IS, 520hp	
Max road speed (km/h)	37	37	
Armour (mm)			
hull front	90	90	
hull side	60	60	
hull rear	60	60	
hull roof	30	30	
hull floor	20	20	

Guns and Armaments

Soviet tank types each mounted a particular main armament, ranging from the 45mm 20K to the 152.4mm ML-20S; the following tables give essential data on these weapons, such as penetration capability and muzzle speed.

Main Guns	20K		KT	PS-3	L-10	L-11	
Calibre (mm)	45		76.2	76.2	76.2	76.2	
Rate of Fire (per minute)	7–12		5	11–12	12	-	
Muzzle Speed (m/s)	757		370	530	558	612	
Shell Weight (kg)	–		6.5	6.5	6.5	6.5	
Guarantee	80%		–	–	–	–	
Penetration (mm)							
100m	43 (60°)	51 (90°)	–	–	–	55 (60°)	66 (90°)
300m	36 (60°)	43 (90°)	–	–	–	–	–
500m	31 (60°)	38 (90°)	31 (90°)	–	50 (90°)	49 (60°)	60 (90°)
1000m	28 (60°)	35 (90°)	28 (90°)	–	–	43 (60°)	52 (90°)
1500m	–	–	–	–	–	38 (60°)	46 (90°)
2000m	–	–	–	–	–	34 (60°)	41 (90°)

Main Guns	F-32		F-34		ZiS-5		ZiS-3	
Calibre (mm)	45		76.2		76.2		76.2	
Rate of Fire (per minute)	3–8		3–8		4–8		6	
Muzzle Speed (m/s)	612		662		662		662	
Shell Weight (kg)	6.6		6.5		6.5		6.5	
Guarantee	-		20–80% calibre		Under calibre		Under calibre	
Penetration (mm)								
100m	43 (60°)	51 (90°)	89–74 (60°)	94–86 (90°)	95 (60°)	120 (90°)	95 (60°)	120 (90°)
300m	36 (60°)	43 (90°)	82–69 (60°)	90–81 (90°)	75 (60°)	90 (90°)	75 (60°)	90 (90°)
500m	31 (60°)	38 (90°)	76–62 (60°)	84–75 (90°)	50 (60°)	60 (90°)	50 (60°)	60 (90°)
1000m	28 (60°)	35 (90°)	71–55 (60°)	78–68 (90°)	–	–	–	–
1500m	–		55–48 (60°)	69–62 (90°)	–	–	–	–
2000m	–		–	–	–	–	–	–

Main Guns	ZIS-2		ZIS-4		D-5T		ZiS-S-53	
Calibre (mm)	57		57		85		85	
Rate of Fire (per minute)	15–25		6–10		5–8		6–10	
Muzzle Speed (m/s)	990		995		792		792	
Shell Weight (kg)	3.15		3.15		9.2		9.2	
Guarantee	–		20-80%		80%		80%	
Penetration (mm)								
500m	85 (60°)	100 (90°)	89–83 (60°)	105 (90°)	90 (60°)	105 (90°)	90 (60°)	105 (90°)
1000m	70 (60°)	90 (90°)	85–78 (60°)	98 (90°)	85 (60°)	100 (90°)	85 (60°)	100 (90°)
1500m	65 (60°)	70 (90°)	79–73 (60°)	90 (90°)	78 (60°)	115 (90°)	78 (60°)	115 (90°)
2000m	55 (60°)	65 (90°)	–	–	72 (60°)	110 (90°)	72 (60°)	110 (90°)

Main Guns	D-10S		M-30	A-19S	D-25T	
Calibre (mm)	100		122	122	122	
Rate of Fire (per minute)	4–6		–	2.2	1.2–2.5	
Muzzle Speed (m/s)	880		–	–	780	
Shell Weight (kg)	15.88		–	25	–	
Guarantee	80%		Cumulative	–	80%	
Penetration (mm)						
500m	125 (60°)	155 (90°)		–	122 (60°)	152 (90°)
1000m	110 (60°)	135 (90°)	100	–	115 (60°)	142 (90°)
1500m	95 (60°)	115 (90°)		–	107 (60°)	133 (90°)
2000m	87 (60°)	110 (90°)		–	97 (60°)	122 (90°)

Main Guns	M-10T	ML-20S
Calibre (mm)	152.4	152.4
Rate of Fire (per minute)	–	2.3
Muzzle Speed (m/s)	436	600
Shell Weight (kg)	40	43.6
Guarantee	HE	–
Penetration (mm)		
500m	–	–
1000m	–	125
1500m	–	–
2000m	–	–

Machine Guns	DT	DShK	TNSh
Calibre (mm)	7.62	12.7	20
Length Calibres	–	-	82.4
Practical Fire Rate	80–125	80	200
Muzzle Speed (m/s)	840	830–850	817
Angle	–	–	60°
Guarantee	–	–	20%
Penetration (mm)			
50m	–	–	25
100m	–	–	18
500m	–	16	–

Flamethrowers		
Flamethrower KS-24	–	30–40m
Flamethrower KS-25	By 9 litres	45–53m
Flamethrower ATO-41	By 10 litres	60–65m / 90–100m
Flamethrower ATO-42	By 10 litres	70m / 100–130m

An IS-2 heavy tank advances in support of Red Army infantry, Ukraine, summer 1944. The long 122mm (4.8in) D-25T gun with its distinctive overhang was an extremely powerful weapon.

Causes of T-34 Tank Losses (as a percentage)

Period	20mm (0.78in)	37mm short (1.5in)	50mm long (2in)	50mm (2in)	75mm (2.9in)	88mm (3.5in)	105mm (4.1in)	128mm (5in)	AT rocket	Unknown
Up to September 1942	4.7	10.07	7.5	54.3	10.1	3.4	2.9	–	–	7.1
Stalingrad operation	–	–	25.6	26.5	12.1	7.8	–	–	–	28.0
Central Front, Operation 1943	–	–	10.5	23.0	40.5	26.0	–	–	–	–
1st Byelorussian Front, Jun–Sep 44	–	–	–	–	39.0	38.0	–	–	9.0	14.0
1st Byelorussian Front, Jan–Mar 45	–	–	–	–	29.0	64.0	–	1.0	5.5	0.5
1st Ukrainian Front, Jan–Mar 45	–	–	–	0.5	19.0	71.0	0.6	–	8.9	–
4th Ukrainian Front, Jan–Mar 45	–	–	–	–	25.3	51.5	0.9	–	9.0	13.5
1st Byelorussian Front, Oder–Berlin 1945	–	–	–	1.4	69.2	16.7	–	–	10.5	2.2
2nd Guards Tank Army, Berlin 1945	–	5.4	–	–	36.0	29.0	6.6	–	22.8	–

Combat Ranges of AFV Engagements, 1943–44

(percentage of Soviet AFVs knocked out by range)		
Distance in metres	**75mm gun**	**88mm gun**
100–200	10.0	4.0
200–400	26.1	14.0
400–600	33.5	18.0
800–1000	7.0	13.5
1000–1200	4.5	8.5
1200–1400	3.6	7.6
1400–1600	0.4	2.0
1600–1800	0.4	0.7
1800–2000	0	0.5

T-34/76 Model 1942 tanks await orders to move out on the Central Front, December 1942. The dual hatch configuration led to the Germans dubbing the model 'Mickey Mouse', because of the similarity with the cartoon character.

Soviet Tank Production

Improved Soviet war production was achieved at the expense of civilian living standards – the most thorough application of the principle of total war – and with the help of Lend-Lease supplies. Germany's advantages in good-quality engineering and skilled labour were offset by greater Soviet efficiency in using their often limited resources. Soviet factories were ordered to concentrate on the production of low-cost, low-maintenance AFVs in contrast to German attempts to gain decisive qualitative superiority by producing sophisticated, costly designs such as the Panther. All major Soviet types were incrementally upgraded while simplified and refined manufacturing processes increased production.

TANK PRODUCTION BY TYPE AND YEAR	1941	1942	1943	1944	1945	TOTAL
Light Tanks						
T-40	41	181	–	–	–	222
T-50	48	15	–	–	–	63
T-60	1818	4474	–	–	–	6292
T-70	–	4883	3343	–	–	8226
T-80	–	–	120	–	–	120
Sub-total	**1907**	**9553**	**3463**	**-**	**-**	**14,923**
Medium Tanks						
T-34	3014	12,553	15,529	2995	–	34,091
T-34/85	–	–	283	11,778	7230	19,291
T-44	–	–	–	–	200	200
Sub-total	**3014**	**12,553**	**15,812**	**14,773**	**7430**	**53,582**
Heavy Tanks						
KV-1	1121	1,753	–	–	–	2874
KV-2	232	–	–	–	–	232
KV-1S	–	780	452	–	–	1232
KV-85	–	–	130	–	–	130
IS-2	–	–	102	2252	1500	3854
Sub-total	**1353**	**2533**	**684**	**2252**	**5001**	**8322**
Total Tanks	**6274**	**24,639**	**19,959**	**17,025**	**8930**	**76,827**
Assault Guns						
SU-76	–	26	1928	7155	3562	12,671
SU-122	–	25	630	493	–	1148
SU-85	–	–	750	1300	–	2050
SU-100	–	–	–	500	1175	1675
SU-152	–	–	704	–	–	704
ISU-122/ISU-152	–	–	35	2510	1530	4075
Sub-total	**-**	**51**	**4047**	**11,958**	**6267**	**22,323**
Total AFVs	**6274**	**24,690**	**24,006**	**28,983**	**15,197**	**99,150**

Soviet Tank Strength (monthly, all theatres)

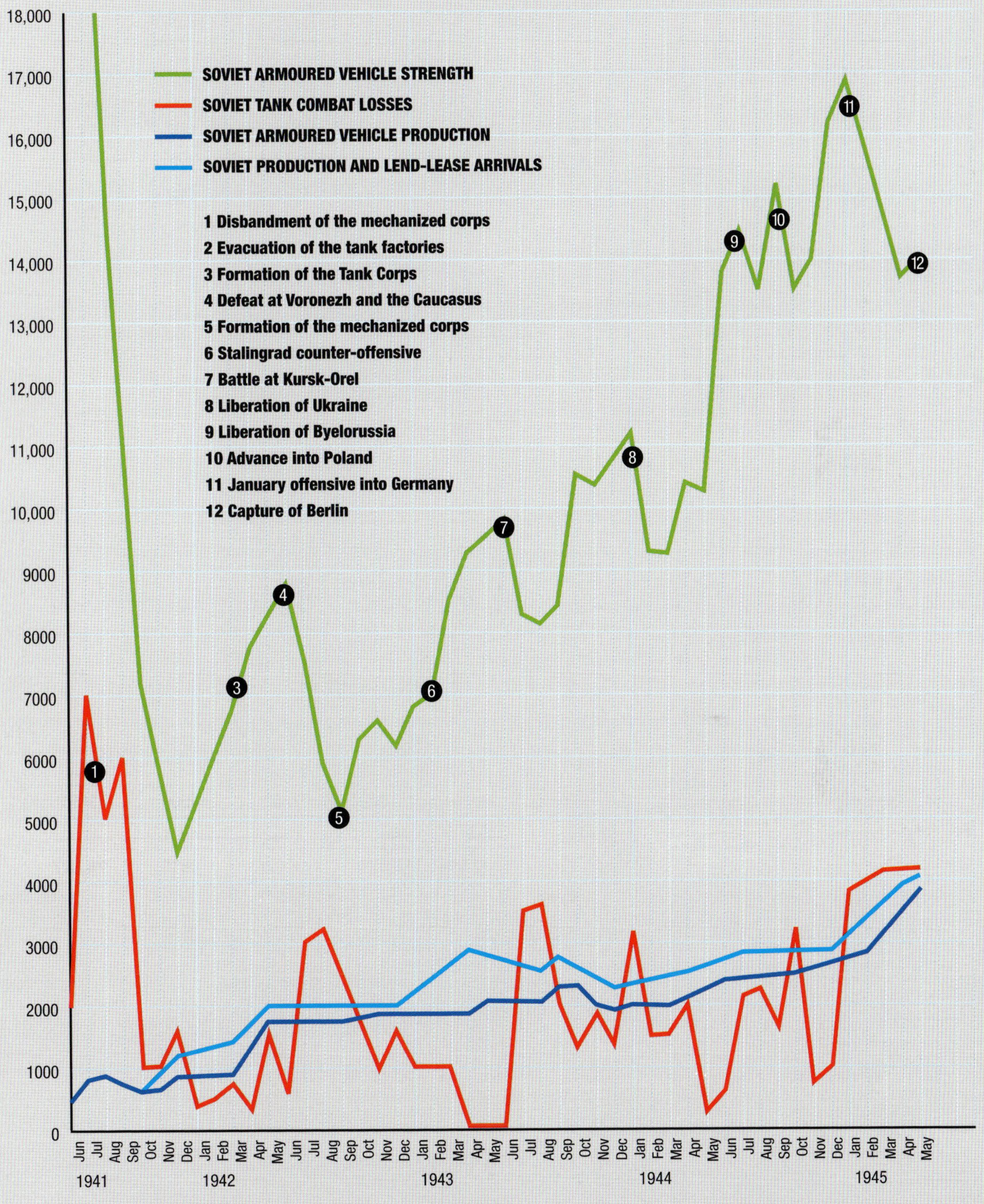

Soviet Automotive Production, 1941–45

	1941	1942	1943	1944	1945	1946
Automobiles	3980	2567	2546	5382	4995	19,470
Trucks	116,169	30,947	45,545	53,467	68,548	314,676
Buses	4027	1462	1175	1700	1114	9478
Total	**124,176**	**34,976**	**49,266**	**60,549**	**74,657**	**343,624**

Lend-Lease Arrivals

The total of Lend-Lease AFVs supplied to the Red Army equalled approximately 16 per cent of Soviet wartime tank production and 12 per cent of self-propelled gun production.

British & Canadian AFVs, Lend-Lease (1941–45)	Sent	Lost	Arrived
Brit Inf Tk Mk II Matilda	1084		
– Matilda Mk III	113		113
– Matilda Mk IV	915	221	694
– Matilda Mk IV CS	156	31	126
Brit Inf Tk Mk III Valentine	2394		
– Valentine Mk II	161	25	136
– Valentine Mk III	346	–	346
– Valentine Mk IV	520	71	559
– Valentine Mk V	340	113	227
– Valentine Mk IX	836	18	818
– Valentine Mk X	74	8	66
– Valentine Bridgelayer	25	–	25
Brit Inf Tk Mk IV Churchill	301		
– Churchill Mk II	45	19	26
– Churchill Mk III	151	24	127
– Churchill Mk IV	105	–	105
Cromwell	6	–	6
Tetrarch	20	–	20
Universal Carriers	1212	not known	–
Canadian Valentine Mk VII	1388	180	1208
Canadian Universal Carriers	1348	not known	–
Total Tanks	**5193**	**710**	**4483**
Total Universal Carriers	**2560**	**224**	**2336**
Total AFVs	**7753**	**934**	**6819**

American AFVs, lend-lease (1942–45)	Sent	Lost	Arrived
M3/M3A1 Stuart	1676	–	–
M5 Stuart	5	–	–
M24 Chaffee	2	–	–
Total Light Tanks	**1682**	**443**	**1239**
M3 Lee medium tank	1386	–	–
M4A2 Sherman (75mm/2.9in)	2007	–	–
M4A2 Sherman (76mm/3in)	2095	–	–
Total Medium Tanks	**5374**	**417**	**4957**
M26	1	–	–
M31B2 ARV	115	–	–
M15A1 MGMC SP AA	100	–	–
M17 MGMC SPG AA	1000	–	–
T48 SPG (SU-57)	650	–	–
M18 tank destroyer	5	–	–
M10 3in GMC TD	52	–	–
M2 halftrack	342	–	–
M3 halftrack	2	–	–
M5 halftrack	421	–	–
M9 halftrack	413	–	–
Total Halftracks	**1158**	**54**	**1104**
Universal Carrier T16	96	–	–
Total AFVs	**8310**	**914**	**7396**

A column of British-supplied Matilda II tanks moves through the Lvov region of Ukraine in the summer of 1944. Although armed with a puny 40mm (1.45in) 2-pounder gun, the tank's thick armour gave the type good combat survivability.

Soft Vehicles, Lend-Lease	**1941**	**1942**	**1943**	**1944**	**1945**	**Total**
Towing Vehicles					–	–
Studebaker	–	3800	34,800	56,400	19,200	114,200
GM	–	1400	4900	400	–	6700
International	–	900	1800	100	300	3100
Chevrolet	–	2700	13,100	25,100	6800	47,700
Ford	–	400	500	–	100	1000
Dodge 3/4	–	–	4300	10,700	4600	19,600
Trucks	520				71	559
Ford-6	–	7600	18,600	29,000	5800	61,000
Dodge 11/2 ton	–	8000	1500	100	–	9600
Dodge 3 ton	–	–	1400	300	–	1700
Bedford	–	1100	–	–	–	1100
Ford Marmon	200	300	–	–	–	500
Austin	200	300	–	–	–	500
Light Vehicles	151				24	127
Willys	–	5400	13,900	14,300	6200	39,800
Bantam	–	500	100	–	–	600
Chevrolet	–	–	–	–	200	200
Special Auto	1212				–	–
Dodge Staff Car	–	–	–	100	100	200
Ford Amphibian	–	–	–	1900	300	2200
GM Amphibian	–	–	–	–	300	300
Trailer	–	–	–	600	200	800
Mack Diesel	–	–	–	–	900	900
Other	–	–	200	300	–	500
Total	**400**	**32,400**	**95,100**	**139,300**	**45,000**	**312,200**

Bibliography

Books:

Alexander, Arthur J. *Armor Development in the Soviet Union and the United States.* Santa Monica, CA: RAND, 1976.
Glantz, David M. *Soviet Military Operational Art: In Pursuit of Deep Battle.* London: Frank Cass, 1991.
Glantz, David M. *When Titans Clashed: How the Red Army Stopped Hitler.* Lawrence, KS: University Press of Kansas, 1995.
Green, Michael. *Russian Armour in the Second World War: Rare Photographs from Wartime Archives.* Barnsley: Pen & Sword, 2013.
Merriam, Ray. *Soviet Heavy Tanks: World War 2 Album.* Bennington, VT: Merriam Press, 1995.
Zaloga, Steven J. and James Grandsen. *Soviet Tanks and Combat Vehicles of World War Two.* London: Weidenfeld & Nicholson, 1984.
Zaloga, Steven J. and Leland S. Ness. *Red Army Handbook, 1939–1945.* Stroud, Gloucester: Sutton Publishing Ltd, 1998.
Zaloga, S., and P. Sarson. *T-34/76 Medium Tank 1941–1945* (New Vanguard No. 9). London: Osprey, 1994.
Zaloga, S., and P. Sarson. *T-34/85 Medium Tank 1944–1994* (New Vanguard No. 20). London: Osprey, 1996.

Websites:

wio.ru/tank/tth.htm
www.globalsecurity.org/military/world/russia/tank-3.htm
www.militaryfactory.com/armor/ww2-soviet-tanks.asp
www.missinglynx.com/articles/russia/rpt34.htm - Ray Peterson 'The T-34/85 in WWII: A Closer Look'.
www.o5m6.de/
www.onthewaymodels.com/articles/T-34-85turrets/Identification.htm - 'A Short Guide to 1/72 T-34/85 Turrets in Modelling: Turret Identification.'
www.tanks-encyclopedia.com/
www.wardrawings.be/WW2/Files/1-Vehicles/Allies/2-USSR/
www.wwiivehicles.com/soviet-union/default.asp

Index